Strategic Health Technology Incorporation

Synthesis Lectures on Biomedical Engineering

Editor
John D. Enderle, *University of Connecticut*

Strategic Health Technology Incorporation

Binseng Wang

ISBN: 978-3-031-00511-4   paperback
ISBN: 978-3-031-01639-4   ebook

DOI 10.1007/978-3-031-01639-4

A Publication in the Springer series
SYNTHESIS LECTURES ON BIOMEDICAL ENGINEERING

Lecture #32
Series Editor: John D. Enderle, *University of Connecticut*

Series ISSN
Synthesis Lectures on Biomedical Engineering
Print 1930-0328   Electronic 1930-0336

Strategic Health Technology Incorporation

Binseng Wang
ARAMARK Healthcare Clinical Technology Services

SYNTHESIS LECTURES ON BIOMEDICAL ENGINEERING #32

ABSTRACT

Technology is essential to the delivery of health care but it is still only a tool that needs to be deployed wisely to ensure beneficial outcomes at reasonable costs. Among various categories of health technology, medical equipment has the unique distinction of requiring both high initial investments and costly maintenance during its entire useful life. This characteristic does not, however, imply that medical equipment is more costly than other categories, provided that it is managed properly. The foundation of a sound technology management process is the planning and acquisition of equipment, collectively called technology incorporation. This lecture presents a rational, strategic process for technology incorporation based on experience, some successful and many unsuccessful, accumulated in industrialized and developing countries over the last three decades.

The planning step is focused on establishing a Technology Incorporation Plan (TIP) using data collected from an audit of existing technology, evaluating needs, impacts, costs, and benefits, and consolidating the information collected for decision making. The acquisition step implements TIP by selecting equipment based on technical, regulatory, financial, and supplier considerations, and procuring it using one of the multiple forms of purchasing or agreements with suppliers. This incorporation process is generic enough to be used, with suitable adaptations, for a wide variety of health organizations with different sizes and acuity levels, ranging from health clinics to community hospitals to major teaching hospitals and even to entire health systems. Such a broadly applicable process is possible because it is based on a conceptual framework composed of in-depth analysis of the basic principles that govern each stage of technology lifecycle. Using this incorporation process, successful TIPs have been created and implemented, thereby contributing to the improvement of healthcare services and limiting the associated expenses.

KEYWORDS

medical equipment, healthcare technology, planning and acquisition, hospital capital equipment, technology assessment, strategic planning, clinical engineering, return on investment, total cost of ownership, technology deployment, equipment selection and procurement, alternative procurement methods

Contents

CHAPTER 1

Introduction

Technology is indispensable for the delivery of health services even in the poorest and most remote areas of the world. Drugs, implants, disposable products, and medical equipment are major contributors to the fantastic progress of healthcare in the last 100 years when compared to the preceding thousands of years. Unfortunately, technology also is a significant contributor to the fast and steady rise of healthcare costs (CMS, 2000; Cutler and McClellan, 2001; Rothenberg, 2003; AdvaMed, 2004; Kaiser Family Foundation, 2007). This lecture covers the process of planning and acquiring technology with the goal of maximizing benefits (clinical outcomes and financial returns) and lowering costs (both investment and recurring).

Although health technology encompasses medical and surgical procedures, drugs, biologics, capital and non-capital devices, support systems (e.g., blood banks and clinical laboratories), information system (e.g., medical records), and organizational and managerial systems (see, Appendix A - Glossary), this lecture will focus mainly on medical equipment because it is the least understood and probably the worst managed of all technologies. Apparently, due to its size, high capital investment, and complex and costly life-long maintenance requirements, medical equipment can be easily used as the "poster child" of uncontrollable rise of healthcare expenses, although actually it is not the most costly category of health technology, even among all types of medical devices (Wang et al., 2008). Nevertheless, the principles of the first part of this lecture (technology planning) can be used almost universally for all categories of health technology.

The incorporation process described in this lecture covers not only technologies that the health organization or system purchases but also goods donated, leased, or borrowed, as well as replacement of existing equipment or introduction of brand new technologies. Furthermore, it applies to freestanding institutions as well as systems composed of thousands of hospitals of multiple levels, health centers and community clinics, although the complexity and timeframe are quite different from one case to another.

The need for a rational and systematic process for incorporating medical equipment is evidenced by the large amount of unsuccessful attempts in wealthy, prestigious hospitals and nations, as well as in less resourceful organizations and countries. In developing countries, many governments—and lenders—have become bitterly disappointed in discovering that the equipment purchased through international loans and donated by philanthropic organizations did not bring the health outcomes desired and large amount of acquired equipment lay idle due to lack of funds for consumables and maintenance challenges (WHO, 1987, 1990; WHO guidelines, 2000; Cho, 1988; Bloom, 1989; Uehara, 1989; Wang, 1989; Erinosho, 1991; Coe and Banta, 1992; Temple-Bird, 2000; Quvile, 2001; Wang, 2003; WHA, 2007).

Similar challenges also exist in developed countries, although at a different level. In the United States, healthcare spending per capita has grown almost four times faster than inflation since the 1970's (Modern Healthcare, 2008) and its national health expenditure has exceeded 16% of the gross domestic product (GDP) in 2006 (ACHE, 2008) and is expected to reach 18.7% by 2014 (Modern Healthcare, 2005). A significant portion of the total expenditure and its rapid growth is attributed to health technology (CMS, 2000; Cutler and McClellan, 2001; Rothenberg, 2003; AdvaMed, 2004; Kaiser Family Foundation, 2007). Reports of assessments performed at individual healthcare organization have confirmed that improper incorporation of technology often leads to lackluster outcomes, increased costs, abusive use, and frustrated health managers, users and patients (ECRI, 1989; David and Judd, 1993).

To be fair, health leaders faced numerous challenges in making technology decisions. As shown on Figure 1.1, a wide range of stakeholders apply considerable pressure over the decision makers, presenting convincing arguments from different perspectives. Patients typically have high

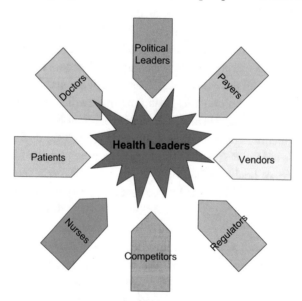

Figure 1.1: Challenges posed by different stakeholders on health leaders in making appropriate decisions on incorporation of medical technology. While each group of stakeholders defend their own interests with little or no consideration for others, the health leaders need to find a way to reconcile these wishes and demands, considering the availability of resources.

hopes that their illnesses can be cured by new technologies. Doctors and nurses are fascinated by new tools, hoping their work will become easier and more effective. Vendors (manufacturers and their distributors) want to sell more products and services. Payers (insurers and their sponsors) want to reduce their costs. Regulators demand compliance with laws, regulations, codes and standards, thinking that safety and quality will follow naturally. Competitive healthcare providers use technology as marketing tools for attracting both clinicians and patients. Political leaders are anxious to fulfill their promises and demonstrate their leadership by opening new facilities and getting their pictures taken next to big, shining new equipment. Health leaders are left with the unenviable job of reconciling all these wishes and demands, looking for a compromise between available resources and desired outcomes.

The best way health leaders can address this enormous challenge is not to let technology incorporation become an *ad-hoc*, subjective process that leaves almost everyone unsatisfied and, most importantly, prevents the organization[1] from fulfilling its mission of providing quality care for its patients. By adopting and gradually fine-tuning a rational and open process for technology incorporation, health leaders can reduce, if not avoid all together, the pitfalls that others have encountered before them. As shown in Figure 1.2, a *Technology Incorporation Plan* can be derived by considering

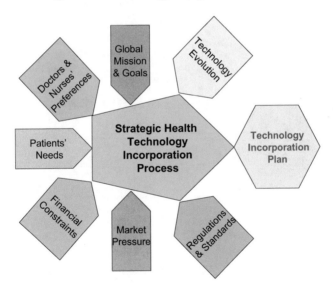

Figure 1.2: The strategic health technology incorporation process is the tool needed by health leaders to combine and reconcile the genuine needs and concerns of all stakeholders and reach a compromise solution that maximizes benefits (clinical outcomes and financial returns) at the lowest possible costs (both investment and recurring).

genuine patient needs, doctors and nurses' preferences, financial constrains, technology evolution provided by vendors, regulatory impositions, competitive market pressure, and global mission and goals, including those coming from political leaders. Obviously, none of the stakeholders will be able

---

[1]Heretofore the term *organization* is used to designate any type and size of healthcare facility, ranging from a single health clinic to a nationwide health system, including hospitals of different sizes and specialties, multi-hospital systems, integrated delivery networks, etc.

to dictate their individual wishes but everyone's input will be considered. By involving all stakeholders and using a multi-disciplinary approach, a compromise solution (preferably by consensus) can be found that is agreeable to most if not all.

For didactic reason, technology planning and technology acquisition are presented as two distinct steps. However, it is important to keep in mind that these two steps are intimately related and the incorporation process as a whole will not be effective and efficient if one of the two steps were conducted without the other one. Furthermore, the material presented is not meant to be a recipe to be followed by the letter, but guidelines that hopefully will provide direction and provoke thinking and understanding of the principles, allowing continual improvement of each organization incorporation process. Before describing the incorporation process, it is necessary to lay the conceptual foundations that will be used throughout the lecture.

CHAPTER 2

Conceptual Framework

2.1 THE ROLE OF TECHNOLOGY

A fundamental concept that should guide the entire incorporation process is that *technology is nothing but a tool.* By itself, technology[1] has little intrinsic value and its outcomes depend on who and how it is used. Like a hammer or saw, it can be used to build a shelter or kill a person, depending on the user's intent and skills. As shown in Figure 2.1, technology is the means through which the health needs and anticipated benefits are fulfilled in the form of impacts on patients, users, infrastructure, and costs. If planned and acquired properly, technology can help health leaders and workers to achieve their goals and objectives of treating and caring for patients in the best and most cost-effective manner. Improperly used, it can hurt people and waste valuable, limited resources.

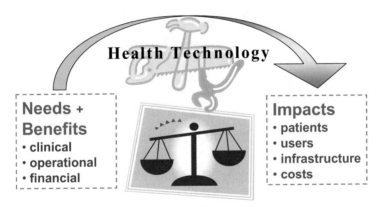

Figure 2.1: Health technology is nothing but a tool that has little intrinsic value but can be invaluable in providing high-quality care in a cost-effective way if used by the right person(s) at the right time and in the right manner. It is, therefore, incumbent on the health leaders to manage technology properly so there is a balance between the needs and desired benefits on one hand, and the impacts (positive and negative) on the other.

2.2 TECHNOLOGY PRODUCTION LIFECYCLE

In managing the technology incorporation process, it is helpful to understand that the technology producers have a different perspective from the healthcare organizations. This is not to say that

[1]Most dictionaries define technology as the application of knowledge for practical purposes.

the former are greedy or exploiting the latter, simply each side has its own goals and objectives. As the blue curve on Figure 2.2 shows, the quantity of equipment produced has a bell like shape as a function of time, starting from its conception (research and development - R&D) and ending with abandonment [see, e.g., Coe and Banta (1992); David and Judd (1993)]. This cycle can be as short as a few years and as long as centuries.

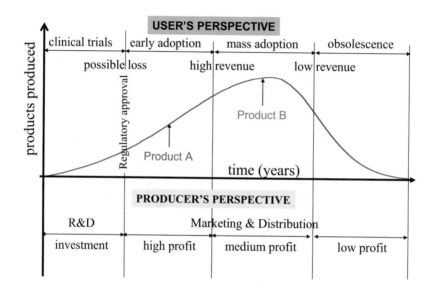

Figure 2.2: Technology production lifecycle from two perspectives, producer's and user's. Producers invest heavily in R&D to gain return later until the product is no longer in demand (obsolescence). Innovative products command premium prices until competitive products are introduced or the market is saturated. Early adopters seek marketing edge and/or high revenue at the risk of not recovering their investments. Late adopters have much lower risks but cannot command premium prices for their services. Product A is typically adopted by large, academic centers, whereas product B is more appropriate for community, non-profit facilities.

During R&D, producers make heavy investments and engage some users in the clinical trials. Once the technology is approved by regulatory agencies for sale, the manufacturers start to market and promote the product while charging a premium price. The early adopters risk their money and reputation hoping to reap high profits and/or gain market share. When the adoption is wide spread, competitive products usually appear and force a reduction in the price for the product. At this stage, users will not be able to charge premium prices as other healthcare organizations will also offer the same services. Finally, when a new technology (represented as product A on Figure 2.2) is launched, the older technology (represented as product B on Figure 2.2) may soon enter into the obsolescence phase, although it may take many years before it is completely displaced and abandoned.

Although the bell-shaped curve applies to most technologies, the exact shape and width can vary significantly from one to another. Some may have a plateau that extends for several years or decades. Instead of putting two or more points on a single curve, it may be necessary to draw two or more curves of different shapes and starting points and compare them. It is also important to remind the reader not to confuse the market life cycle with the management life cycle. The former describes the process by which a technology is invented, developed, marketed and, eventually, abandoned in the market place (Figure 2.2); whereas the latter describes the process by which the technology is incorporated, maintained, managed and, eventually, retired within a health organization or system (Figure 2.3).

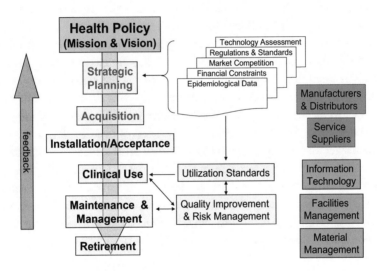

Figure 2.3: Technology management lifecycle within healthcare organizations. Unlike drugs, implants, and disposable devices, equipment needs to be managed from its incorporation (planning and acquisition) until retirement, always guided by the organization's policy (mission and vision). Numerous external inputs are needed for incorporation, including information and services provided by manufacturers, distributors, and service providers. Within the organization, while Clinical Engineering (CE) is the primary steward of technology, it needs to work closely with Information Technology, Facilities Management, and Material Management, in addition to the clinical users. The most important and often neglected segment is the feedback provided by users and CE professionals for future incorporations. Without the benefit of accumulated experience, one is condemned to "reinventing the wheel" eternally.

When evaluating technology for incorporation into an institution or system, a critical question to be asked is "where is this technology in its life cycle?" Is it in clinical trials, early adoption, mass adoption, or obsolescence stage? In general, it is risky to incorporate a technology when it is still being tested and introduced into the market, although the risk may pay off in terms of market competitiveness or higher profit. As a rule of thumb, public and non-profit organizations, with the exception of academic research centers, should avoid incorporating any technology that has not yet reached mass adoption stage. Within any health system, only the highest levels of care should consider early adoption. Primary and secondary care institutions should rely on well-proven technologies. Obviously, no one should incorporate technologies that are clearly being made obsolete by superior ones (but also not discard the obsolete ones unless they are unsafe or no longer economical). Care must be exercised when obsolete technologies are donated (see discussion below). Private and academic institutions need to have access to newer technologies for competition and research purposes, and should be prepared to write off some investments that do not pay off in the future.

After determining where in its life cycle a particular technology is located at the time of planning, planners must determine which competitive and substitutive technologies are available or about to be introduced into the market. Once found, the question to be addressed is how likely and soon these competitive and substitutive technologies will make obsolete the technology under consideration. The difference between competitive and substitutive technologies is in how they address the needs. An example of competitive technologies is the self-capping (or "auto disable") needles versus the needle destruction devices. They compete with each other with the same goal of reducing needle-stick injuries. An example of substitutive technology is drug patches that deliver drugs without injections and, therefore, eliminating the need of needles altogether.

One way to answer all these questions is to review medical literature and perform technology assessment (see details below) using internal resources. Another is to use publications and/or services provided by consulting companies that collect this kind of information (see consulting companies listed in Appendix B). Although these publications and services are often fairly expensive, the savings that can be achieved often compensate the initial investment. Needless to say, although manufacturers are happy to provide information and assistance, their advice is seldom unbiased.

2.3   TECHNOLOGY MANAGEMENT LIFECYCLE

Among health technologies, medical equipment presents some unique challenges that set it apart from others. Unlike drugs, implants, and disposable supplies, medical equipment is not consumed by the patients. That is good news in the sense that it can be reused many times over. The bad news is that it requires continued care, much like vehicles and aircrafts. Actually, there are many similarities between healthcare organizations and commercial airlines. Both are service entities that use technology and people to achieve their objectives (provide health care and transporting passengers, respectively). The first one uses medical equipment (and other technologies), while airlines use aircrafts (also other technologies). Both kinds of equipment require maintenance and management, not only to conserve the capital investment but to ensure the ability to generate revenue and safeguard their customers (patients and passengers, respectively). Hospitals employ doctors, nurses, etc., while airlines have pilots, flight engineers and attendants, etc. All are highly trained professionals, often licensed by regulatory agencies, who understand that mistakes can lead to injuries and deaths and, thus, cannot be tolerated.

Therefore, it is no surprise that commercial airliners are maintained according to rigorous rules. In the USA, the Federal Aviation Administration (FAA) is responsible for the oversight of aircraft mechanics and how their work is performed. Paradoxically, there are no similar oversight for the maintenance and management of medical equipment. Hospital licensing agencies have often adopted the guidelines recommended by the Centers for Medicare and Medicaid Administration (CMS, State Operations Manual, Publ. #100-07) or deferred to accreditation organizations such as the Joint Commission (2009), American Osteopathic Association (AOA, 2005), or DNV Healthcare Inc. (DNV, 2008). In the USA, almost anyone can claim to be competent and be employed by hospitals to repair medical equipment, calling themselves a biomedical technician or clinical engineer. Fortunately, the leaders of this incipient profession have initiated voluntary certification programs (see ACCE-HTF and ICC in Appendix B) that eventually may lead to a formal licensed profession. Until the profession is legally structured, it has been designated variously as biomedical engineering (typically within American hospitals but not in American universities, where biomedical engineering is viewed as the broad application of engineering in biology and medicine), asset management, technology management, or clinical engineering. In this lecture, the term clinical engineering (CE) will be used heretofore and its professionals referred to as CE professionals.

Even with the absence of specifically license professionals, healthcare organizations still need to manage properly medical equipment to provide care, minimize unnecessary expenditure, and protect patients. Figure 2.3 shows the equipment management lifecycle should start from the mission, vision, goals and objectives defined by the organization's governing body in the form of a health policy (or a set of policies), including strategies and tactics to address specific needs and achieve certain goals. From this policy, technology needs are strategically planned using information available from a variety of sources, including laws and regulations, voluntary standards, clinical guidelines, technology assessment studies, financial constraints, competitive analyses, etc. This is the beginning of the incorporation process, which extends into selection and acquisition, addressed in this lecture. Once the necessary and appropriate equipment is acquired, it will be installed (if necessary) and tested before acceptance and clinical use (to be addressed in future lectures).

After proper user training and qualification, clinical use will be guided by utilization standards, quality improvement goals, and risk management considerations. Periodically, the equipment must be inspected for safety and performance and, if needed, repaired, updated, or upgraded as required by recalls. Finally, when the equipment ceases to be productive, presents unreasonable risks, or is no longer needed, it will be retired (destroyed, exchanged, or transferred to another facility). During the entire life cycle, numerous factors and entities have direct and indirect impact on the quality and longevity of the technology, such as the manufacturers, distributors, service providers, and the environment (facilities and utilities, information technology, and procurement of supplies). One of the most important elements that are often deficient in poorly managed equipment life cycles is the feedback process from all levels back to the initial policy making and planning processes. Without this feedback, it is difficult, if not impossible, to learn from the mistakes made and improve future incorporations.

2.4   HEALTH TECHNOLOGY ASSESSMENT

The incorporation (or the planning portion only) of medical equipment (and other types of health technology) is often confused with or called "health technology assessment." This is unfortunate, as health technology assessment (HTA) is a discipline by itself, with rigorous methodologies and much broader scope and goals than technology incorporation. It also has its own professional association (Health Technology Assessment International - www.htai.org). Furthermore, HTA is one of the essential information sources for a good incorporation process[2].

Goodman (2004) defines HTA as *"the systematic evaluation of properties, effects or other impacts of health technology. The main purpose of HTA is to inform policymaking for technology in health care, where policymaking is used in the broad sense to include decisions made at, e.g., the individual patient level, the level of the health care provider or institution, or at the regional, national, and international levels. HTA may address the direct and intended consequences of technologies as well as their indirect and unintended consequences. HTA is conducted by interdisciplinary groups using explicit analytical frameworks, drawing from a variety of methods."*

The impacts studied by HTA are typically divided into the following classes: (i) safety; (ii) efficacy and/or effectiveness; (iii) economics; and (iv) social, legal, ethical, and/or political implications. Traditionally, safety and efficacy far exceed the other impacts, as it is unnecessary to consider other impacts if a particular technology is found to be unsafe or incapable of improving health. For this reason, the most important source of information for HTA has been randomized, controlled clinical trials (Coe and Banta, 1992). It is then quite obvious that technologies that have not been proven to be safe and effective by HTA studies should not be considered further for incorporation. In addition to safety and efficacy/effectiveness information, it would be quite desirable to have data on the other two classes of impacts for proper incorporation decisions. Unfortunately, these data are hard to find, probably due to challenges in obtaining unbiased data for objective studies.

2.5   TECHNOLOGY DEPLOYMENT COSTS

A financial model made popular by the Gartner Group (www.gartner.com) for the deployment of large information technology (IT) systems is the total cost of ownership (TCO) model. This model highlighted the fact that the initial investment in hardware and software often pales in comparison to the subsequent hardware and software maintenance and upgrades, and especially to the "hidden" cost of user training and continuous learning.

As mentioned before, among health technologies medical equipment has the unique characteristic of a durable good that requires recurring expenses to operate and maintain. This means that it is most inappropriate to make acquisition decisions using solely the purchase price (including the costs of packaging, shipping, insurance, taxes or import duties, and installation when and wherever applicable). This is because the post-acquisition costs of operation, maintenance, administration, and

---

[2]A broader assessment of medical practice is called "comparative effectiveness," which is defined as *"a rigorous evaluation of the impact of different options that are available for treating a given medical condition for a particular set of patients"* (CBO, 2007).

learning typically far exceed those of initial investment. Like IT deployment, the initial investment for incorporating medical equipment represents about 20% of TCO, whereas the costs of the supplies needed for operating the equipment, labor and parts needed for maintenance, administrative costs of managing the capital assets, and training and retraining of users and maintainers easily add up to 80% of TCO (Figure 2.4). The user learning costs are particularly difficult to measure as the time spent learning new systems and training fellow workers increases rapidly with the accelerated introduction of computers and software into medical equipment.

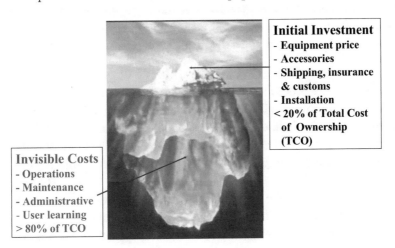

Figure 2.4: The total cost of ownership (TCO) of medical equipment is analogous to an iceberg. The initial investment is only ~20% of TCO, whereas the recurrent costs of operating and maintaining it for its useful life can reach ~80% of TCO. Health leaders aware of this fact can reduce the risk of becoming the next "captain of the Titanic."

TCO can be divided into four categories depending on its relationship with the technology in question (direct or indirect) and whether it is an initial investment or a recurrent expenditure. A fifth category, end-user costs, is segregated from indirect, recurrent costs due to its unique nature. A brief description of each category follows.

2.5.1   DIRECT INVESTMENT

The capital investment directly related to the purchase of equipment includes the free-on-board (FOB) cost of the equipment itself and its accessories. Although called accessories, they are often essential to the operation of the equipment, e.g., probes for ultrasound imaging equipment. To the FOB cost, the vendor typically adds the packaging and shipping costs, known as "shipping and handling – s/h" costs. The price of equipment at the point or port of arrival is known as CIF cost, i.e., cost with insurance and freight. Some states or countries may impose sales tax or import duties on acquisitions, depending on the nature of the purchasing organization (i.e., ownership and/or profit orientation). If customs clearance is needed, it may be necessary to add the fee to customs brokers. Also, preparations for installation and the installation itself must be included in the investment costs. This is especially important for large equipment that may require new construction or refurbishment of an existing building, as well as changes in electrical supplies (e.g., high voltage lines), water, gases, air conditioning, protection against radiation and electromagnetic interference (EMI), etc.

Among the investment costs often overlooked are those associated with the training of users and service personnel, supplemental furniture (e.g., trolleys and stools), and specialized test and calibration equipment required for maintenance. The training sometimes requires travel to another city or country in addition to tuition or fees. If included in the capital investment, these costs can be amortized and properly accounted for in the financial reports.

2.5.2   DIRECT RECURRENT COSTS

Most equipment requires certain supplies for its operation. General supplies include utilities like electricity, water, medical gases, etc., while specialized supplies can be in the form of chemical reagents, films, liquid nitrogen, rechargeable batteries, printer paper, etc. All equipment will eventually need repairs and, sometimes, periodic inspections and/or preventive maintenance (collectively known as schedule maintenance - SM). These services can be provided by in-house CE staff, manufacturer's representative, or independent service organizations. The outside services can be contracted on an as-needed basis (often known as time and material – T&M) or through a comprehensive service contract, or any arrangement in between these two extremes. Typical total maintenance cost ranges from 2-15% of the original acquisition cost, depending on the complexity and exclusivity of the technology, as well as the local support environment. Included in this broad range is the cost of replacement parts, maintenance supplies, and labor. In general, industrialized nations have lower parts and supplies costs, while developing countries have lower labor costs. Finally, the most important recurrent cost directly attributable to the incorporation of technology is the personnel cost, including clinical operators and CE staff. Even though typically only a portion of their time is devoted to the equipment, there is no denial that all equipment requires attention from the users and servicers.

2.5.3   INDIRECT INVESTMENT COSTS

Often not accounted in the incorporation process are the expenses indirectly related to planning and acquisition, such as acquisition of HTA reports, consultant fees, time spent by clinical, technical, and administrative staff, financing costs, purchasing and importation paperwork (e.g., certificate of need application, customs clearance, etc.), custom warehousing, etc.

2.5.4   INDIRECT RECURRENT COSTS

After the equipment is installed and operating, there are still numerous indirect costs that need to be accounted. Some examples include archival of additional patient records, warehousing of supplies and replacement parts, employee protection (e.g., radiation dosimeters, face masks, etc.), disposal of hazardous materials (biological and chemicals contaminants, radioactive waste, etc.), additional liability insurance coverage (for high-risk equipment and associated procedures), etc.

2.5.5   "HIDDEN"/END-USER COSTS

The most often ignored costs are those related to the users because these costs are in the form of staff time and, therefore, are difficult to quantify. This fact was first noticed and gained attention in IT with the rapid introduction of personal computers (PCs) into corporations. The Gartner Group (http://www.gartner.com) has published numerous studies showing the importance of these costs, which they call "end-user operations." Due to the progressive incorporation of microprocessors and computers into medical equipment, clinical users are experiencing similar types of challenges and are spending significant portions of their time in learning and using the equipment, thereby reducing their overall productivity. Typical "hidden"/end-user costs include: self-learning and peer training, customization of user interface and reports, data management (archiving, recovery, etc.), applications development (e.g., setting up specific sequences of MR imaging), and troubleshooting (attempts in finding problems before calling for support).

CHAPTER 3

The Incorporation Process

Figure 3.1 shows the health technology incorporation process as composed of two steps: planning and acquisition, each of which is composed of several sub-processes. As discussed before (Figure 2.3), these two processes do not exist by themselves. They are part of the lifecycle of medical equipment management within a health organization, guided by the policy, mission, vision, and strategies defined by the organization, to be followed by the installation and acceptance process within technology management, and continually refined with feedback from all the subsequent processes.

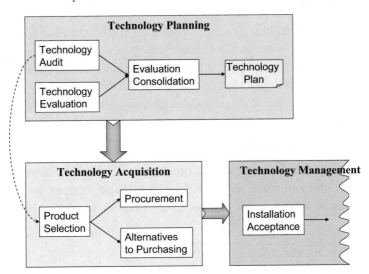

Figure 3.1: The technology incorporation process divided into two major steps: planning and acquisition. Within each step, there are several sub-processes that need to be performed in order to bring the desired technology into the organization for it to be installed (if needed) and accepted, before being used on a patient. The division into steps and sub-processes is arbitrary but convenient for didactic purposes. Furthermore, there are important links among the sub-processes, such as the one indicated by dashed line between technology audit and product selection for standardization purposes that should not be ignored during the implementation of an incorporation project.

Within the planning section, the needs, impacts, costs, and benefits of technology are evaluated after an audit of the existing technical resources is performed. The data collected by the audit and evaluations are consolidated and converted into a *Technology Incorporation Plan* (TIP) that will guide future capital investments and current maintenance strategies. In the acquisition section, the TIP is implemented by the selection of products that are most appropriate for the particular application and environment and, subsequently, by the procurement of the selected products. Whenever possible and advantageous, alternatives to purchasing (e.g., leasing, "committed supplies purchase agreement," and revenue-sharing arrangements) should be considered. Each of these sub-processes will be discussed in detail below.

Due to the wide variety and range of costs of medical equipment and of health organizations, not all parts of the process is necessary for all cases. Therefore, readers must use discretion to determine which portions, if any, of the process is appropriate for each application. For example, it would be inappropriate to apply a comprehensive process for purchasing a defibrillator for a single hospital or clinical instruments needed by a single primary health center. On the other hand, the entire process may be necessary for planning, selecting, and acquiring a magnetic resonance imaging (MRI) system for a single organization or hundreds of defibrillators or clinical instrument kits for a system. Likewise, the replacement of existing equipment by products based on similar technologies (e.g., surgical lights) does not require an extensive research when compared with the deployment of new technologies with which few persons within the system or organization are familiar (e.g., surgical robots). However, the general principles presented below are always applicable and have been proven to be useful in developed countries [see, e.g., David and Judd (1993); ECRI (1997); Taylor and Jackson (2005)] and developing countries [see, e.g., Coe and Chi (1991a,b)].

3.1   GOALS AND OBJECTIVES

As stated in the beginning of this lecture, the goal of strategic technology incorporation is to maximize benefits (clinical outcomes and financial returns) while minimizing costs (both investment and recurring). The specific objectives can vary from one organization to another but typically include some of the following:

- Improve patient outcomes and satisfaction

- Increase access of patients to care

- Widen the coverage of patient populations and geographical areas

- Reduce risks to patients, clinicians, and environment

- Maintain (or improve) organization's market/leadership position

- Balance clinical needs and staff wishes against available capital resources

- Adopt proactive planning to address long-term needs, reducing emergency acquisitions

- Reduce total cost of ownership (TCO)

- Offer more learning opportunities for clinicians and students (if academically affiliated)

- Maintain (or increase) standardization to improve efficiency and reduce risks

- Comply with group purchasing organization (GPO) contracts

Each organization should add its own objectives to the list above according to its unique circumstances and challenges.

3.2   INCORPORATION RESOURCES

Before describing the principles and techniques that are used for technology planning and acquisition, a discussion of the resources needed for the entire incorporation is made to emphasize its importance. Without appropriate resources, it is not possible to conduct incorporation properly and the results will likely to be disappointing. Even with sufficient resources, they need to be structured and empowered in the proper manner to function efficiently and effectively.

As mentioned above, while the principles and techniques discussed here are generic enough to be applied to a wide range of organizations, the reader must decide how to scale it properly for a particular organization in order to match the resources invested with the actual need. Furthermore, it is likely that even within a well-endowed organization, not every incorporation project requires all the resources and the complete set of tools and methods. In order not to overdo the incorporation, it is advisable to divide technology into groups according to its purchasing cost (or better yet, its TCO) and increase resources as the cost increases.

Like many endeavors, the most critical resource necessary is human resources. Two common mistakes are either excessive centralization of decisions or too much reliance on subject matter experts (SMEs). The former leaves all the decisions to the chief executive officer (CEO) or the board chairperson, thus creating political problems related to lack of transparency, favoritism, and subjectivity. The latter often lacks credibility and widespread support, sometimes aggravated by the lack of broader vision.

The most effective way to combat subjective demands ("wants/wishes" of fancy equipment that would serve as toys or status symbols for certain individuals) and political pressures (promises made in election campaigns) is to establish a multidisciplinary team—the *Technology Incorporation Committee* (TIC)—composed of representatives of the main stakeholders, i.e., medicine, nursing, pharmacy, nutrition, administration (including finance), and support services (including facilities, information technology, clinical engineering, and material management). The peer review process requires rational and scientific-based justifications that will stand up to criticisms.

Whenever possible, TIC should be composed by the highest ranking person of each department because the presence of the senior leadership provides more credibility and weight to its decision and recommendations. On the other hand, the involvement of important executives like the chief medical officer (CMO), chief nursing officer (CNO), chief finance officer (CFO), chief operations officer (COO), vice president of support services, etc. may make it difficult to schedule meetings and assign homework. A possible compromise is to ask each executive to appoint a representative to ensure broad representation and support for the decisions and recommendations.

The primary duty of TIC is to interpret the organization policies, establish priorities, and marshal the necessary resources to implement technology incorporation. Depending on the size of the organization and, therefore, of the TIC, it could perform many of the planning and acquisition activities itself. More likely than not, it will have to create task forces to address specific subjects. The TIC's role would then be of reviewing and discussing the conclusions and recommendations made by each task force and decide on what to recommend to the CEO or Board of Directors. The task forces would then be composed of SMEs in their respective professions capable of providing perspectives from their respective areas of expertise.

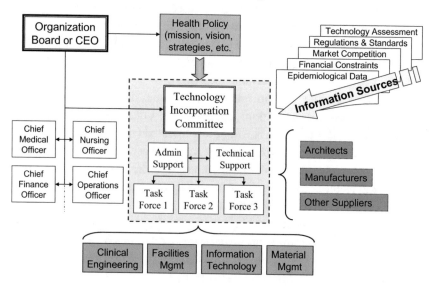

Figure 3.2: Internal and external resources needed for technology incorporation. Within the organization, a Technology Incorporation Committee (TIC) should be assembled with the participation of key stakeholders of the highest possible level (e.g., the chief officers shown on the left side). TIC will be supported by administrative and technical staff, and created task forces to address specific types of technologies or equipment. External resources can be in the form of information and vendors and consultants.

As shown on Figure 3.2, regardless of the composition of the TIC and task forces, they will need to collect information from both internal and external sources. Some examples of internal sources and respective types of information to be collected are:

- current users: safety, effectiveness, ease of use, training, etc.

- clinical engineering: reliability, safety, maintainability, etc.

- facilities management: utilities requirements, environmental impact, etc.

- information technology: networking issues, software support, etc.

- material management: supplies, accessories, alternative vendors, etc.

From outside of the organization, the following sources and types of information should be considered:

- health information clearinghouses: epidemiological data, reimbursement consideration, regulations and standards, market competition, financial issues, HTA reports, etc.

- manufacturers: product specifications, financial terms, installation and operational requirements, warranty, post-sale support, etc.

- architects and civil engineers: infrastructure requirements and impacts, codes and regulations, etc.

- other suppliers: ancillary equipment and furniture, alternative sources of supplies and services, etc.

Finally, TIC and its task forces will need appropriate support both administratively and technically. Most of these support needs can be fulfilled by existing internal resources but should be properly accounted for, so it will be easier to measure the indirect investment costs mentioned earlier.

For large health systems, it is critical to adapt TIC structure and relationship with the components of the system. For example, if the system has been decentralized by granting certain autonomy to its components, a central TIC still is necessary, but its focus should be on large capital investments that have broader impact than a single system component. For example, the cold chain necessary for system-wide vaccination campaigns or the emergency medical services (and the associated equipment) for disasters should be planned centrally by coordinating regional efforts. Lower cost and geographically-limited-impact technologies should be left for each component to plan and acquire with its own resources. Having said that, it is worth pointing out that the local decision makers are likely to need support from a team of highly specialized and experienced multidisciplinary staff. Many health systems, especially those in developing countries, have a shortage of such personnel and it is impossible to expect all local decision makers to assemble their own teams of multidisciplinary experts. An example of such a central planning team that supports a decentralized health system was described by Wang (1990) for Sao Paulo state in Brazil and a similar approach was later adopted in South Africa (2001).

3.3 STRATEGIC HEALTH TECHNOLOGY PLANNING

The primary goal of technology planning is to establish a TIP, i.e., which and how many pieces of equipment are needed for a certain period of time (e.g., 3-5 years). The TIP should be tightly linked with the policies and strategies (i.e., mission, vision, goals and objectives) defined for that organization by their respective decision makers.

3.3.1   TECHNOLOGY AUDIT

If TIP is created for a new organization or wholesale replacement of organization (e.g., those destroyed by major disasters), then there is no need to know what already exists. Most plans, however, are for improving or expanding existing infrastructure. In this case, it is essential to know the quantity and quality (i.e., operating conditions) of what exists, including the reasons for premature deterioration or damage.

Numerous commercial software packages are available for inventory and management of medical equipment (Cohen, 2003). However, a simple database can be built and used without necessarily acquiring commercial software. One of the main challenges in data collection is the nomenclature for equipment, as the same piece can be known to different users with multiple names. Furthermore, the frequent mergers and acquisitions of companies among manufacturers make the identification and grouping of equipment somewhat challenging. For this reason, it is essential that the inventory is managed by people with good command of the technology (known as biomedical engineering technicians – BMETs or clinical engineers).

On the other hand, caution must be taken not to make the inventory excessively detailed (e.g., incorporating each and every subassembly or module of a system, and accessories). This is because the information collected is ephemeral and can become outdated even before the inventory is completed. Unless the information is to be used also for other purposes (e.g., equipment maintenance), the inventory should be limited to basic information such as: equipment name, manufacturer, model, serial number, asset identification, year of production or acquisition, current operational status, current location, major failures in the last 3 years and respective causes.

3.3.2   TECHNOLOGY EVALUATION

Four types of evaluation should be performed for technology planning: need, impact, costs, and benefits. Each of these four evaluations is discussed below.

3.3.2.1   Evaluation of Need

The first step of planning is to assess the real need for technology. It is critical to differentiate need from wishes. Many people would like to drive a Lexus or even a Porsche but do they really need such a sophisticated car? Can they support one even if they were given a luxury vehicle without having to pay for it?

A more rational and systematic approach is not to ask which types and quantity of equipment each clinician, department, or hospital desires, but to ask "which diseases or conditions do you want to address?" or "what are the policies or priorities defined by the health decision makers? "Then, ask the healthcare professionals what they need to address the diseases or conditions, or implement the defined policies or priorities in the safest and most cost-effective manner.

The answers to these questions are often available from local or international literature in the form of clinical studies, standards of care or, in the case of newer technologies, HTA reports. Some health systems have published manuals or standards that provide lists of instruments, equipment, and other infrastructure needs for health clinics, rural hospitals, and community hospitals [e.g.,

Mexico (1999); Brazil (1994)]. More complex situations, however, demand investigation by a team of multidisciplinary SMEs, taking into consideration the level of care (primary, secondary, tertiary, etc.) and local realities. In this situation, TIC should create a task force focused on the particular clinical area or geographical region to determine the means needed to provide prevention, diagnosis, therapy, and/or rehabilitation.

The task force should limit its attention to safety and clinical effectiveness of the available technology, as the issues related to impacts and financial sustainability will be evaluated later. In addition to the safety and effectiveness for the patients, the task force needs to broaden its scope to the users, the general population, and the environment. For example, the older radiotherapy equipment based on radioactive substances such as cesium and cobalt requires not only protection for the patients and the clinical operators but also posed significant safety challenges to the people and environment, as well as terrorism concerns. Some HTA reports completed between 1990 and 1999 by the Agency for Health Care Policy and Research (now called Agency for Healthcare Research and Quality) can be found on the website of the National Library of Medicine (http://www.ncbi.nlm.nih.gov/books/bv.fcgi?rid=hstat6.part.38807). These reports are somewhat outdated but can serve as starting points for research on more up-to-date information.

While reviewing data submitted by manufacturers, the task force needs to insist in obtaining recall information from all countries in which the products are distributed. No matter how careful the manufacturer designs and builds the equipment, there is always a possibility that problems will be detected only months or years after the product is launched into the market. To correct these unforeseen problems, manufacturers are required by regulatory authorities to issue safety alerts and recalls that require the owners or users to modify or dispose the unsafe product. While no one should expect the manufacturers to be perfect at their first attempts, past recall history can provide valuable insight into not only how well each manufacturer manages risks but also the risks that are inherent to certain technologies and need to be managed by the healthcare organizations and practitioners.

Once it is established that safety and effectiveness of a particular equipment clearly outstrips potential risks, the next step is to determine the quantity of equipment needed, considering the epidemiological data, population to be covered, geographical distances (actually transportation times as discussed later), and equipment capacity (or per case usage duration). Obviously, any calculation is an estimate based on available data and can be significant off the target if the data are not reliable. Nonetheless, one can only learn by taking some risks and make adjustments learning from mistakes. Again, the "feedback" arrow shown on Figure 2.3 serves as a reminder of the need to learn and improve.

3.3.2.2 Evaluation of Impact

As mentioned before, equipment differs from other types of health technology in its long-lasting, multi-dimensional impact on healthcare organizations. Before a piece of equipment is incorporated into the health service, a study must be performed on the impacts that this equipment will have on the service directly and indirectly. One of the reasons identified in many organizations, especially in developing countries (Project HOPE, 1982; WHO, 1987, 1990; WHA, 2007), that causes under-utilized equipment is the failure to anticipate and plan for its impacts. These impacts can be divided into three main groups: infrastructure, users, and maintenance.

Impact on Infrastructure   Among the many impacts any new equipment will have are those that affect the infrastructure such as physical space (size and foot print), weight, mounting requirements, environment requirements (temperature, humidity, light, etc.), utilities (electricity, water, gases, etc.), and protection (against radiation, laser, etc.). Incorporating equipment without understanding its infrastructure requirements will lead to headaches and considerable additional expenses. These impacts are especially important for rural areas and developing countries where stable sources of power, abundant supply of clean water, and controlled environment (temperature and humidity) are often not available.

Typically, the infrastructure impact data can be obtained from the manufacturers, often known as "installation requirements." For large systems (e.g., MRI), most manufacturers provide architectural blueprints and layouts of supporting structures.

Impact on Users   Another equally important impact that must be understood and planned is the training needed by physicians, nurses, therapists, etc. to use the equipment in a safe and effective manner. Often users are required to perform comprehensive pre-use checks, wear personal protective equipment, perform emergency intervention on equipment failures, and understand the limitations of the technology. According to data collected and analyzed by the Joint Commission (2004), clinician orientation and training is the second leading root cause of all sentinel events (i.e., incidents that caused or could have caused harm to patients) at ~57%. Although the actual amount of medical equipment related sentinel events caused by user orientation and training was not published, it is likely that a significant portion of the equipment incidents are caused by inadequate training. Fortunately, the amount of equipment related sentinel events has been fairly low [<2% of all sentinel events each year per Joint Commission (2008)].

Besides the initial training (known as "inservice"), it is essential to plan for periodic user training for refreshing the concepts and to address the inevitable personnel turnovers. Even if retraining by manufacturer's representative is guaranteed, it is critical to require detailed user manuals that provide step-by-step instructions and precautions, so refreshers can be provided by on-site educators and assisted by clinical engineering staff.

Impact on Maintenance   As one of the basic tenets of healthcare is patient safety, maintenance of medical equipment goes far beyond the traditional roles of preventive and corrective maintenance common to industrial equipment. Scheduled safety and performance inspections (SPIs) are critical tasks performed to detect hidden and potential failures that are undetectable by the clinical users but can cause injury or death to patients. For example, an audible alarm on a ventilator can fail without anyone knowing until it is activated by a patient condition or equipment failure. Another example would be an intra-cardiac pressure monitor could leak minute amount of electrical current that can shock and fibrillate a patient's heart, without the clinicians ever feeling anything (even when they are not wearing gloves).

Highly trained and qualified professionals (clinical engineers and BMETs) are required to perform these inspections and maintenance, even during the warranty period. Ideally, these people should receive "factory" training together with the manufacturer's own service staff. However, due to high travel costs and tuition charges, such training may not be possible except for the most expensive and sophisticated equipment. In addition to qualified technical staff, it is often necessary to acquire specialized test and measurement equipment, including software if needed. Finally, service manuals with detailed mechanical diagrams, electronic circuits, software description, and parts lists are indispensable for continuous support. For organizations and systems located in remote areas or outside of developed countries, special attention must be paid to the access of maintenance parts. Even if financial resources are available, the time consumed for shipping and custom clearance may substantially delay the repair of mission-critical equipment to the point of jeopardizing patient flow and even safety.

In essence, the decision makers must understand that the three basic elements (known as the "*trinity*") of maintenance are: qualified personnel, service documentation, and spare parts. The lack of any one of these three elements will affect the availability, reliability, and safety of the equipment.

3.3.2.3 Evaluation of Costs

As shown on Figure 2.4, the majority of expenses associated with equipment incorporation is incurred after the equipment is acquired and installed. As the five main categories of costs are already covered in detail in Section 2.5 above, each of them will be treated only briefly below.

Direct Investment Costs   In addition to the costs associated with the equipment purchase, such as equipment and accessories, packaging, shipping, taxes, customs, etc., this category includes costs related to training (users and maintenance professionals), maintenance tools and equipment, infrastructure work (derived from the impact evaluation above), and personal protective equipment, etc.

Direct Recurrent Costs   This category includes all the resources needed to operate the equipment, such as:

- Human: dedicated operator, time of non-dedicated clinicians, maintenance professionals, and administrative support;

- Material: single-use and single-patient-use consumables, supplies, chemicals, utilities, replacement parts, etc.;

- External resources: maintenance provided by manufacturers and their distributors or third-party companies, testing, calibration and/or protection verification, software upgrades, etc.

Indirect Investment Costs   Costs not directly attributable to the incorporation but, nevertheless, needed are included in this category. Examples include acquisition of HTA reports, consultant fees, time spent by clinical, technical, and administrative staff, financing costs, purchasing and importation paperwork (e.g., certificate of need application, customs clearance, etc.), custom warehousing, etc.

Indirect Recurrent Costs   Often equipment creates additional demands on related services and, thus, generates indirect recurrent costs. Examples include archival of additional patient records, warehousing of supplies and replacement parts, employee protection (e.g., radiation dosimeters, face masks, etc.), disposal of hazardous materials (biological and chemicals contaminants, radioactive waste, etc.), additional liability insurance coverage (for high-risk equipment and associated procedures), etc.

"Hidden"/End-User Costs   This is the most difficult to estimate of all categories, because it involves time spent by a large amount of individuals who are impacted by the technology as they try to perform their duties. Since there is no easy way to track the time spent on self-learning and peer training, customization of user interface and reports, data management, applications development, etc., it can only be estimated grossly. It should be clear to the decision makers that these costs can be reduced by proactively investing in training and support services, which also can contribute to increased efficiency.

3.3.2.4   Evaluation of Benefits

While all the costs of incorporating equipment are being evaluated, it is also necessary to evaluate the potential benefits, direct and indirect, that the equipment being considered will bring to the organization or system. Three categories of benefits are discussed below: clinical, financial, and indirect.

Clinical Benefits   The clinical benefits are those associated with patients and population who will receive preventive care, be diagnosed, treated, or rehabilitated. Examples of typical benefits are: reduction in deaths or disease (or gains in life expectancy or quality of life), increase in effectiveness, increase in productivity (or reduction in procedure time for prevention, diagnosis, therapeutics, or rehabilitation), and increase in reliability. Whenever possible, it would be most advantageous to quantify these clinical benefits (e.g., using disability adjusted life years saved – DALYs) so comparisons can be made between similar products and with competitive technologies.

Financial Benefits Financial benefits associated with incorporation of new equipment or replacement of older technologies often are in the form of: increased revenues (e.g., due to higher productivity, higher reimbursement rates, etc.) for organizations allowed to charge their patients, and decrease expenditures (e.g., due to lower supplies and maintenance costs, lower liability risks, shorter length of stay (LOS), reduction in referrals to other institutions, etc.). Again, careful calculation of the projected values will be helpful for quantitative comparisons later. Depending on the relevance, it may be worthwhile to perform detailed financial analysis such as return on investment (ROI), feasibility or break-even analysis, and tax implications.

Indirect Benefits Planners should not ignore many indirect benefits useful not only for comparing products and technologies later, but also for assisting the ultimate decision of acquiring or not the equipment. Some examples of indirect benefits are: increased user productivity due to the "satisfaction" of working with newer technologies (an important psychological factor) or better designed user interface, additional revenues for other procedures (e.g., additional surgeries after diagnosis made by some sophisticated imaging equipment under the fee-for-service or diagnostic-related-group models), increased patient throughput, broader coverage, and improved access. In addition, depending on the relevance, it may be worthwhile to include marketing and competitive benefits such as recruitment of prestigious physicians and surgeons, market competition, and higher reimbursement rates due to higher ranking by insurance companies or government authorities.

3.3.3 EVALUATION CONSOLIDATION

One of the biggest challenges of technology evaluation is how to consolidate the data in a meaningful way so good decisions can be made. Many of the evaluation results are not quantitative and these evaluations are in widely different dimensions (e.g., financial versus clinical or social).

Since the ultimate goal is to produce a TIP that lists the equipment to be acquired by order of priority, an approach that has been adopted by many organizations and systems is to rank all the prospective incorporations in such a way that the highest priorities will surface to the top, while the less important ones will stay at lower levels. Depending on the availability of capital, the higher ranked needs will be acquired first, then the middle ranked ones, and so on until the money runs out.

Figure 3.3 shows a spreadsheet used for ranking some fictitious equipment, using the evaluation variables discussed above. For each piece of equipment scores were attributed in the range of -5 to $+5$ (to clearly differentiate between the negative and positive impact of each piece of equipment, respectively) and the total score was computed by first averaging all the scores in each category and then applying a weighting factor (in this example, set at 20% for need, impact, and costs, and 40% for benefits). This list is then sorted by the total score in descending manner, so the highest scored item is shown at the top. The scores shown on this figure were purposefully assigned in such a way to demonstrate a few basic concepts:

ITEM	EQUIPMENT BEING EVALUATED	NEED EVALUATION	IMPACT EVALUATION			COSTS EVALUATION			BENEFITS EVALUATION			TOTAL SCORE	INVESTMENT COST (US$)	CUMULATIVE COST (US$)
			Infrastructure	Users	Maintenance	Investment	Recurrent	Users	Clinical	Financial	Indirect			
1	Video-endoscopy for lower GI	3.5	0.0	-1.5	-2.0	-1.0	-1.0	-2.0	3.0	5.0	3.0	1.7	$20,000	$20,000
2	YAG surgical laser	3.0	0.0	2.0	-1.5	-2.5	-2.0	-2.0	2.0	4.0	3.0	1.4	$65,000	$85,000
3	Cardiac ultrasound system	3.0	0.0	-3.0	-1.5	-3.0	-1.5	-2.0	3.0	2.0	4.0	1.1	$60,000	$145,000
4	Endoscope washer and disinfection	4.5	-0.5	-1.0	-1.0	-1.0	-2.0	-2.0	2.0	1.0	1.0	1.1	$12,000	$157,000
5	Automated chemistry analyzer	2.5	-2.0	-1.5	-3.0	-3.0	-3.0	-2.0	5.0	3.0	2.0	0.9	$250,000	$407,000
6	Surgical light	2.5	-1.0	1.0	-1.0	-2.0	0.0	0.0	2.0	1.0	0.0	0.7	$25,000	$432,000
7	Intra-aortic balloon pump	3.0	-1.0	-1.0	-2.5	-2.0	-2.0	-1.0	3.0	1.5	1.0	0.7	$50,000	$482,000
8	250 general purpose infusion pumps	2.5	0.0	2.0	-1.0	-3.5	-2.0	-2.0	3.0	1.0	0.0	0.6	$750,000	$1,232,000
9	Video-endoscopy for upper GI	1.5	0.0	-1.0	-1.5	-1.0	-1.0	-2.0	1.0	1.0	2.0	0.4	$20,000	$1,252,000
10	Second CT scanner (64 slice)	2.0	-4.0	-3.0	0.0	-5.0	-3.5	-3.0	4.0	2.0	5.0	0.4	$1,000,000	$2,252,000
11	Phaco-emulsifier for eye surgery	-2.0	0.0	-2.0	-2.0	-1.0	-1.0	-3.0	2.0	2.0	3.0	-0.1	$55,000	$2,307,000
12	Surgical table	-1.0	-1.0	0.0	0.0	-2.0	0.0	0.0	1.0	1.0	0.0	-0.1	$35,000	$2,342,000

Figure 3.3: Example of evaluation consolidation using ranking by attributed scores. The color highlighted cells show the highest and lowest scores of each evaluation category. The gray highlighted cells on the last column show equipment that will not be covered by current budget.

(1) The piece of equipment considered most advantageous for marketing differentiation purposes (the new, 64-slice CT scanner) was ranked fairly low in spite of its high Indirect Benefit score (5.0), due to the high impacts on infrastructure, user training, and maintenance, as well as costs for investment, recurrent operations (X-ray tubes), and impacts on user learning and picture archiving.

(2) The item with the highest clinical benefit score (automated chemistry analyzer) also did not rank at the top for similar reasons. However, it still ranked higher than the CT scanner.

(3) The endoscope washer was the most needed piece of equipment and it rightfully reached a fairly high rank, enough to ensure its acquisition even if the total capital budge were limited to $0.5 million.

(4) The highest ranked piece of equipment was a video-endoscopy system for lower gastro-intestinal application. The main reason being the expected financial return, but the other factors were also in its favor.

The right two columns on Figure 3.3 show the individual direct investment costs (see Section 3.3.2.3 above) and the cumulative costs of all equipment ranked above that particular line item. The last column allows TIC to visualize easily how their budget is affected by their evaluation. It is not unusual that after seeing the results that TIC decides to rescore all items again to try to fit better the equipment into the available budget. In this example, it is assumed that the budget is limited to $1.5 million, so only the top 9 items can be incorporated; the remaining items are shown in gray, meaning that they will need to be considered at a later opportunity.

In essence, this example illustrates the need to consider all types of impacts, costs and benefits before making a decision. Even when a particular piece of equipment reached the top score (priority) in a single category, it may not win the top spot at the end due to other factors that need to be considered. It is not uncommon that some surprises come out from this kind of exercise.

The ranking is obviously nothing but a quantification (or hierarchical ordering) of subjective judgments. However, it is easier to sort a long list by number rather than by multiple non-quantitative attributes (e.g., grades from A to F). As long as the scores are assigned systematically by all the participants (even though each may have his/her own personal bias), the end result is still valid and useful.

It should also be pointed out that the evaluation criteria discussed above are not the only ones that should be used. Each organization or system should select those that are applicable and discard the superfluous ones, as well add those that are missing. Furthermore, in the example shown on Figure 3.3, the total score was computed with a weighting factor that needs to be evaluated for appropriateness and adjusted if needed.

3.3.4 PLAN REVIEW AND APPROVAL

After TIC is satisfied with the results of its evaluation and ranking, the TIP is submitted to the organization or system senior leadership for review and approval. Besides seeking their blessing, this is also an opportunity to advocate for more resources or suggest new or changes in strategy, using the data collected during the evaluation process. It is possible that some of the information (e.g., HTA reports, new regulations, market competition, etc.) was not available to the senior leaders when they originally laid out the organization's strategic plan.

TIC should not be surprised to see their TIP changed or even rejected by the organization or system senior leadership. This is possible because the senior leaders often have a much broader view of the challenges and expectations. Ideally, these challenges and expectations should have been carefully communicated to TIC but sometimes there are swift and sudden changes that occurred while TIC is working on its evaluations. This does not mean that their efforts were wasted or the TIP is useless. As President Eisenhower said "*Plans are worthless but planning is everything*" (Eisenhower, 1957), the value of a detailed and careful planning process is to promote deeper understanding of the underlying principles, variables and hypotheses, so that it becomes easier to revise the plan once the results are collected and the need for changes becomes evident. TIC can quickly revise the TIP using the planning process it adopted to take into consideration the new challenges and expectations, and thus generating a revised TIP that will be satisfactory to the senior leadership.

3.4   STRATEGIC HEALTH TECHNOLOGY ACQUISITION

Once the TIP is reviewed and approved by the organization or system senior leadership, its implementation cannot be relegated to administrative staff without supervision and participation of TIC itself or, at least, its task forces. This is because there are numerous critical decisions to be made in the acquisition process as described below. These decisions require persons with not only solid knowledge but also often with clear understanding of the mission and goals to be accomplished with these acquisitions.

Acquisition can be divided into two sub-processes, selection and procurement, even though it is possible to do both at the same time. The advantage of separating them is to decrease the impact of price and allow more time to review all the relevant dimensions before deciding how to weight different dimensions in addition to price. In other words, the selection process will help define how to value the non-monetary aspects like what was accomplished before in non-financial evaluations in planning.

3.4.1   SELECTION

Four dimensions should be considered in the selection process: technical, regulatory, financial, and supplier. Although these four dimensions are applicable to most medical equipment, their relative values could vary significantly from one technology to another, as the relevance of each dimension could be quite different from one to another. For example, supplier qualification and reputation is critical for sophisticated equipment that is likely to require maintenance and software support contracts during its entire life, but much less important for basic equipment that can be supported by in-house or independent, third party engineers and technicians.

It cannot be overemphasized the need of not skipping or cutting short this sub-process. Without good selection, there is a very high likelihood that the acquisition will lead to problems like the purchase of excessively sophisticated equipment, cheap products that are not reliable or robust for intensive use, low priced products with expensive supplies (i.e., high TCO), unsatisfactory local support in spite of good products, etc. Furthermore, this "homework" will provide valuable tools for acquisition decisions, such as bid evaluation or direct negotiations with supplier.

3.4.1.1 Technical Considerations

Although it is not difficult to compare competitive products, the challenge is to determine the characteristics that are essential to ensure its safe and effective use, with the least amount of excessive sophistication. In addition to functional specifications, attention must be given to reliability, maintainability, and productivity. Good starting points are the *Healthcare Product Comparison System* produced by ECRI Institute (https://www.ecri.org/Products/Pages/hpcs.aspx?sub=Capital%20Equipment) and the reports published by MD Buyline (http://www.mdbuyline.com). Although neither report covers all the medical equipment available on the market and neither company is able to keep this comparison always up-to-date because of the continuous introduction of new products worldwide, the characteristics listed in their reports are usually enough as the starting point. Local and special requirements should be added as needed. In addition to technical specifications, attention should be given also to issues such as standardization (see below) and compatibility with existing technologies (e.g., communication standard for digital image archiving systems).

3.4.1.2 Regulatory Considerations

The basic objective is to determine which regulatory requirements all products must comply in order to be sold in the institution's country or state. In addition to the national regulatory registration or approval (e.g., FDA or CE Mark), it is often worthwhile to ask the suppliers to provide information with regard to testing required by state healthcare licensing, worker protection, and fire safety agencies. For example, some states within the US have incorporated the NFPA 99 standards by reference and, thus, the medical equipment is required to have been tested by one of the Nationally Recognized Testing Laboratories (e.g., Canadian Standards Association - CSA and Underwriter Laboratories - UL).

In addition to regulatory requirements, it is worthwhile to review voluntary safety and quality standards [e.g., ISO 14971 (2000) and 9001 standards (2000)] that are often adopted by some manufacturers before being required by law. Obviously, no regulation or standard can provide total absence of risk and maximum quality. However, the more scrutiny a product or a production process has received, the more likely it is less risky or has better quality than others that have been less rigorously examined.

3.4.1.3 Financial Considerations

As mentioned before, it is critical to have a clear idea what are the financial impacts a particular technology will bring to the institution or system. Likewise, different products often have different financial impacts. As in the case of technical comparison, the objective here is to determine the cost elements that will allow discrimination of the competitive products that are acceptable in terms of technical specifications and regulatory compliance. Fortunately, with the wide availability of computerized spreadsheets, it is easy to build a total cost of ownership (TCO) model that will allow not only comparison of competitive products but also specify what cost information the suppliers must provide in their bids or quotations.

3.4.1.4   Supplier Considerations

Acquiring a product, especially a piece of equipment that will last for 5-20 years, is almost equivalent to a "marriage" between the supplier and the customer for that period. It is, therefore, essential for each "partner" to evaluate each other well before "making the vows." In the case of organizations or systems located far from the producers, the supplier includes the manufacturer and its local distributor, which may be a subsidiary or an independent company. The supplier's reputation and the institution's prior experience with the supplier deserve special consideration. Unfortunately, it is difficult to evaluate reputation and past performance objectively and quantitatively. Furthermore, this may not be allowed in a bidding required for public institutions. However, buyers can require performance guarantees in the form of holding a fraction of the payment or deposit a performance bond with a bank until all responsibilities are fulfilled. Serious violations, such as fraud and corruption, can be punished by excluding the supplier from all future biddings.

3.4.2   PROCUREMENT

The goal of procurement is to acquire the needed technology or product that meets the legal, regulatory, and technical specifications, provides the lowest TCO, and satisfies other requirements such as after-sales support, domestic preference, etc. The acquisition may be accomplished through purchasing, leasing, agreements with supplier, and other non-traditional methods. There is not enough space to describe each method in detail here, so only a brief discussion of the advantages and disadvantages of each method and their respective special challenges and issues. The principles and issues discussed here apply to the acquisition of a single piece or numerous pieces of equipment, for a single or multiple locations, new or used equipment, turnkey projects, as well as decentralized systems; however, the process should be scaled to match the volume and cost of goods to be acquired.

3.4.2.1   Purchasing

The most common method of acquisition is purchasing, i.e., gaining the title to the equipment in exchange for money (with or without financing). It is typically the most cost-effective method in the long run and is favored by lending organizations because it allows them to seize the property in case of delinquency. Depending on the accounting rules and tax laws, it also allows private organizations to reduce taxes through amortization. There are numerous ways of conducting the purchase and the most common ones are reviewed below.

Single-Stage Open Bidding   This is the method used by most public organizations as it offers, in principle, the lowest cost and greatest efficiency, fair opportunity for all interested parties, and transparency in the process. As the name implies, the entire process is made in one stage, i.e., all the bids are opened and judged without further interactions between the buyer and bidders. However, unless the process and all the associated documents are carefully prepared and scrutinized, it is relatively easy to defeat its basic goals. For example, the technical specifications can be written to favor a particular brand and model over competitive products without being obvious to untrained persons. For this reason, the success of single-stage open bidding depends heavily on the experience, dedication, and competency of people involved in the entire process. Large biddings can easily require six or more months to conclude, assuming that there are few legal challenges and disputes.

The most serious disadvantage of open bidding is that it is difficult to adopt any decision criteria other than the lowest acquisition price or, at best, TCO. This is because any criteria that combines technical specifications, quality of product and support, and TCO can be challenged legally for being too subjective. So it is not unusual that open biddings often end up buying the cheapest products that do not offer desirable quality or TCO in the long run.

Multiple Stage Bidding   In the case of turnkey projects (e.g., building and equipping an entire hospital) or the purchasing of sophisticated equipment that requires construction or extensive infrastructure preparation, it is often prudent to conduct a two or more stage bidding process. In the first stage, bids are solicited to provide only a conceptual design or performance specifications, with the proviso that changes in technical specifications and execution schedule may be introduced later. The initial bids are reviewed by a multidisciplinary team, which will try to combine the best features of each proposal to draft a set of final specifications to be published for the second stage. The bidders are then required to submit a revised, final technical proposal with firm pricing. Rarely a third stage is necessary, but this may be required if significant changes are needed, the second stage ended up in a tie, or the initial bidding was challenged and had to be voided. This approach is even more time consuming than the single-stage bidding process but allows the buyer the opportunity to use the expertise of the bidders to its own advantage. Therefore, multi-stage bidding should only be used when the complexity and high value justify the time and effort required. It is important to note that this method is distinct from the prequalification method used in the request-for-proposal (RFP) method described below. In the latter, the request is sent only to those who have known and proven capabilities and resources, whereas multi-stage bidding is open to all those who wish to participate.

Request for Proposal (or Limited Bidding)   As mentioned above, instead of opening the bidding process to all who may be interested, a request for proposal (RFP) may be sent only to those who have known and proven capabilities and resources. This method, also known as "limited bidding," is used by most private (for-profit or non-profit) hospitals but is also acceptable even for international loans when the contract values are small or when there is only a limited number of suppliers. Sometimes prequalification (often called "request for information" – RFI) may be necessary to ensure that the

RFP is sent only to those who are qualified. Prequalification may be conducted by asking potential bidders to submit information regarding their (a) experience on similar contracts, (b) capabilities with respect to personnel, equipment, and other resources, and (c) financial status. RFP is preferable to open bidding due to less paperwork and shorter implementation but caution must be exercised to reduce the risks of bias or ignorance. Again, a multidisciplinary approach can help reduce these risks.

Shopping   This is the method used to buy readily available off-the-shelf goods or commodities that are inexpensive. Typically, three or more price quotations are obtained from suppliers to assure competitive prices. Shopping may be conducted domestically or internationally. Some lenders have specific requirements on the maximum values that can be purchased locally and how many suppliers from how many countries must be consulted. Shopping is simpler and faster than both bidding and RFP but needs to be used judiciously to prevent abuse. Both RFP and shopping are useful tools for decentralized health systems that wish to allow local and regional decision makers to acquire low-cost products for immediate use. An example of how price and geographical impact was used in a decentralized system to allow certain items to be purchased locally or regionally using RFP and shopping has been described by Wang (1990).

Direct Purchase   Buying directly from a specific vendor is typically allowed only for private organizations, although sometimes public organizations can use this method to standardize equipment or when the required equipment is proprietary and obtainable only from one source. Direct purchasing can be used to develop long-term relationships with certain suppliers and thus allowing prompt access to newer technologies and gain better support from the suppliers.

Challenges in Purchasing   In implementing one or more of the purchasing processes described above, the following challenges are likely to be encountered:

Technical Specifications   Experience has shown that good and unbiased technical specifications are difficult to produce. Excessively detailed specifications can severely restrict the number of suppliers, while superficial specifications will allow poor-quality products to win due to their lower cost. The following guidance is provided by the World Bank (1999): "Specifications shall be based on relevant characteristics and/or performance requirements. References to brand names, catalog numbers, or similar classifications shall be avoided. If it is necessary to quote a brand name or catalog number of a particular manufacturer to clarify an otherwise incomplete specification, the words "or equivalent" shall be added after such reference. The specification shall permit the acceptance of offers for goods which have similar characteristics and which provide performance at least substantially equivalent to those specified." Unfortunately, there is no single universally accepted technical specification library anywhere in the world. Each organization or system needs to find a way to establish a database of nomenclature and specifications for its planning and acquisition using a multidisciplinary team. An example of such an effort can be found in Sao Paulo state, Brazil, when it implemented the Metropolitan Health Program financed by the World Bank (Wang and Candido, 1989). Another example is the Cuadro Básico developed by the Mexican Ministry of Health (Mexico, 1999). Organizations that do not have an established multidisciplinary team or are starting such a team should examine, in addition to the brochures provided by the suppliers, two publications from ECRI Institute, Healthcare Product Comparison System and Health Devices. The former compares the features of numerous types of medical equipment without testing them, while the latter compares a smaller variety of medical equipment after extensive laboratory tests. Neither one is likely to have all the equipment types or the brands and models, but they provide enough details for clinicians and engineers to start building their own comparison tables and, from there, derive the technical specification that are included in the bidding process.

Financial Evaluation   The majority of bid invitations require the bidders to propose their goods with prices that include shipment (CIF or CIP). This gives the impression that the bids must also be judged solely on price rather than the total cost of ownership (TCO). As discussed before (Figure 2.4) the post-purchase expenses of operating and maintaining a piece of equipment is frequently 2-5 times the purchase price. Furthermore, it is quite common that the product with the lowest purchase price will have the highest TCO over its entire life. Actually, as discussed below, many manufacturers are willing to "give away" their equipment in exchange for a commitment to purchase supplies for a certain period of time (see consumable-purchase agreement purchasing alternative below). For this reason, most if not all bids should be evaluated financially using their TCOs.

Warranty   Often included in the technical specifications, warranty is an important feature that needs to be carefully managed. Some manufacturers are now proposing and many buyers are eager in accepting the idea of "extended warranty." This approach is simply the assimilation of future service contracts (i.e., recurrent costs) into the purchasing price and, obviously, inflating the capital investment. Although extended warranties could be helpful in the initial incorporation of sophisticated equipment for which the organization does not have adequately trained CE professionals they tend to perpetuate the vicious circle of dependence on suppliers. If decision makers ignore training the CE professionals when planning new acquisitions, the organization has no option but to commit to extended warranties in new purchases and sign service contracts at whatever price the supplier demands when the warranty expires.

Involvement of Users   The involvement of users in equipment purchasing is so critical for both care efficiency and patient safety that the Joint Commission started to require their (and CE) participation in the accreditation standards for American hospitals (Joint Commission, 2009). In addition, equipment acquired without active user participation is likely to be under-utilized. The participation of users not only makes them feel part of the process—and, therefore, reduces the risk of rejection of the products purchased—but also helps to increase the transparency in the process. The participation of users should be engaged as early as possible. In the planning process, users can help define more precisely the needs and impacts. In the selection process, users can help established the features that they deem critical. These features, if found reasonable by their peers, can then be included in the technical specifications for bidding. Later, users should be invited to evaluate the bids, both technically and financially. This participation helps to ensure they are convinced that the product purchased has the desired specifications and also the lowest TCO.

3.4.2.2 Group Purchasing

Group purchasing is a method of increasing the buyers' bargaining power by combining purchases from several organizations or even countries. This method is particularly attractive for small organizations and countries that otherwise do not have enough volume to wrestle discount from the suppliers. In the United States, for example, the majority of drugs and disposable devices, as well as a substantial portion of equipment, purchased by hospitals is made through group purchasing organizations (GPOs) because of the substantial discount GPOs can obtain. Another example is the joint purchase of essential drugs and vaccines by Latin American countries through the Pan-American Health Organization (PAHO/WHO), known as the Revolving Fund for Vaccine Procurement.

The primary disadvantage of group purchasing is the lack of freedom to select brands and models that are needed for certain special environments or address cultural requirements. In order to build up the volume, GPOs will only contract with a few suppliers, thereby severely limiting the variety of goods available. GPOs are also difficult to manage because they depend on their members to comply with the decision of the majority and also be prompt on their payments. Otherwise, GPOs will quickly loose their credibility and, thus, bargaining power.

3.4.2.3 Turnkey Projects

Turnkey projects are those when a single contractor or a consortium of companies is hired to carry out a very complex and time-consuming project, e.g., building and equipping an entire hospital or dozens of health centers, one-of-a-kind sophisticated equipment that requires design, adaptation, construction or extensive infrastructure changes. As noted before, turnkey projects often require multiple stage bidding. The bidders are typically required to quote the price of the final product installed at the designated site, including all the associated expenses, including transportation and insurance, installation and commissioning, duties, taxes, etc.

The challenge of managing turnkey project procurement is not limited to the proper conduction of the bidding process. Too often, the lack of involvement of the future users in the architectural design, planning and selection of equipment, and training of operators and maintenance personnel dooms the project from the beginning. Even when the project is well intentioned, funded, and carried out, substandard results can happen because future operational issues were not properly considered in the planning stage (Uehara, 1989).

In general, turnkey projects should be managed as if they were conducted by the institution itself, i.e., the institution needs to assemble a multidisciplinary team that follows the project from conception to reality. A one-to-one counterpart system should be set up, i.e., each contractor's manager should have an institution counterpart who participates in the process as an observer and intervenes whenever the institutions interests could be compromised. All major decisions, e.g., selection of the equipment and definition of the technical specifications, must be agreed and signed off by the institution's representatives before they are implemented. One of the expected outcomes of this counterpart system is the training of the institution's employees for future projects.

On the other hand, turnkey projects do offer some significant advantages. The contractor typically has more experience and better coordination, and can act as a neutral party to resolve political disputes. Also, many hidden administrative costs are unforeseen in the initial planning of large, complex projects but will be covered by the contractor in a turnkey project.

To ensure success, many details need to be included in the agreement with the contractor. These include issues such as complete transfer of documentation (blue prints and computer-aided-designs), training of users and maintenance personnel, warranty for building and equipment, technical support after project completion, and performance security.

3.4.3   ALTERNATIVES TO PURCHASING
The traditional method of acquisition through purchasing could not be the most cost-effective method in the following situations, individually or in combination:

- When the life cycle of technology is very short (see Figure 2.2), so the buyer may end up with obsolete equipment very soon;

- When investment capital is very difficult to obtain or expensive (i.e., the interest rate is very high);

- When local tax laws favor operational expenses instead of amortization of capital investment;

- When the costs of operation and maintenance are much higher than the cost of initial investment;

Numerous alternatives to purchasing have, therefore, been developed and the most common ones are discussed below.

3.4.3.1   Lease
In accounting vocabulary, lease is defined as an agreement between the owner of equipment (lessor) and another person (lessee) by which the lessor agrees to give possession of the property to the lessee in exchange for periodic payments. During the lease period, the lessee is responsible for keeping the equipment in good operational conditions, including the performance of maintenance and recall updates. At the end of the lease period, the lessee may gain ownership of the equipment, have the right to buy it for a preset price, or simply return it to the lessor, depending on the terms of the lease agreement.

Although the legal definition of lease is almost identical in every country, the financial reporting rules can vary significantly. In the United States, the standards issued by the Financial Accounting Standards Board (FASB, 2008) require certain leases should be reported by the lessee as asset acquisition (i.e., "capitalized"); however, the same leases are treated solely as operational expense in other countries. Depending how leases are reported financially, it could be advantageous when compared to purchase. Even if leases have to be capitalized, they could still be advantageous if there is severe limitation in investment capital or the organization is very concerned about rapid obsolescence. Otherwise, lease is in general more expensive in the long run when compared to purchase. Healthcare organizations also need to remember that they need to have good credit standing in order to obtain reasonable leasing terms. Otherwise, the high interest rate may make the lease much less attractive than purchase.

In order to reduce their administrative costs and financial risks, lessors are reluctant to offer lease with periods shorter than 3-5 years. This means that the healthcare organization is required to keep the equipment for at least that amount of time. The combination of lease length and cost means that equipment leases should be planned just as well as any purchase.

3.4.3.2   Rental

Rental is a form of lease that works well for short periods of time (typically <1-2 years but can be as short as a few days) and for mobile equipment (to avoid installation costs). Legally, it is not different from leasing but due to the short commitment, rental is typically considered an "operational lease" and, therefore, an expense and not a capitalized asset in the United States (FASB, 2008). However, the accounting laws and rules may be different in other countries. Rental could be an advantageous approach for institutions and countries whose health insurance reimbursement fully covers direct expenses.

In addition to financial advantage, rental is particularly useful: (i) for covering shortages due to unpredictable patient census variations, especially for therapeutic and life-support technologies, when transferring patients from one location to another is not a good option; (ii) when the institution is not capable or interested in assuming responsibility for maintenance and safety issues, including recall updates; and (iii) if the institution has difficulties in managing downtime and/or utilization, resulting in too many pieces of equipment idle or too few available at critical times. The role of a rental organization is basically to optimize the utilization of limited capital resources in a certain geographical area when the institutions themselves are not capable or interested in managing shared resources (Wang, Sloane and Patel, 2001). This organization can be a government agency or a for-profit business.

3.4.3.3   Consumable-Purchase Agreement

As mentioned above, sometimes the supplier is willing to lend the healthcare organization or system a piece of equipment if the latter commits to purchasing certain amounts of consumable (supplies) per year for a number of years. Furthermore, the supplier assumes the responsibility of providing user training, preventive and corrective maintenance, upgrade, recalls, and replacement of the unit after certain number of years. Examples of this practice are external renal dialysis equipment, clinical laboratory analyzers, and pulse oximeters with disposable probes.

Known as "reagent contract" in the United States (due to its extensive practice in clinical laboratory) and "comodato" in Latin American countries, this kind of agreement is actually a type of lease in legal terms. However, since nothing is paid for the use of the equipment, there is no need to capitalize the expense for accounting purposes.

Consumable-purchase agreements are very convenient for the healthcare organization or system because there is no capital outlay and maintenance headaches. On the other hand, the TCO could be considerably higher than purchasing the equipment and consumables, as the vendor is obviously embedding all these costs into the price of the consumable (supplies). Furthermore, since this is not considered a capital expenditure, it frequently does not require public bidding or is subject to other types of scrutiny. This opens a dangerous loophole that can be exploited by unscrupulous vendors and buyers. Even when bidding is conducted, only vendors who have control of the supplies are qualified to participate, thereby reducing the number of bidders. Organizations need to recognize that this kind of agreement as a form of technology incorporation and manage this process carefully to reduce its expenditures and enhance transparency.

3.4.3.4 Revenue Sharing Agreement

Another method of incorporation without capital outlay is revenue sharing. In this case, the vendor agrees to place a piece of equipment in an organization that agrees to use it to provide services and collect revenue. The two parties share the revenue after deducting the capital depreciation and operating expenses. Again, in legal terms, this is also a type of lease, the vendor being the lessor and the organization, the lessee.

Often a minimum payment must be guaranteed by the lessee to keep the equipment. In this case, the vendor selects and specifies the equipment in common agreement with the organization, while the latter determines outcomes and, therefore, revenue. Both have interests that the joint venture works well so both can be profitable. This arrangement works well for high cost procedures that are reimbursed by insurance and require sophisticated equipment that the institution does not have the investment capital to acquire. The supplier gains the opportunity to "sell" another piece of equipment that would otherwise not be sold. Vendors seldom offer this opportunity to public institutions knowing it would be difficult to manage the utilization and reap profit. In some countries, public institutions are also forbidden by law to enter into profit sharing arrangements with private companies.

3.4.3.5 Donations

At first glance, donation seems to be the least costly alternative to incorporate technology. Unfortunately, the iceberg (Figure 2.4) "hits home" very soon, i.e., the cost of operating and maintaining the equipment may quickly become unbearable for the recipient. The operational and maintenance cost is, however, not the only reason that the recipient needs to plan for donations, as the costs of shipping, installation, and staff training can also be formidable (although the first two costs are sometimes covered by the donor). Some recipients have the misconception that they should just grab whatever is offered and throw out whatever is not needed or usable. This approach not only can strain the relationship with the donors but also the cost of proper disposal of unwanted equipment can be high or even dangerous (e.g., disposal of a radioactive cobalt therapy equipment).

Donors also have made numerous mistakes in the past. Some donors are only interested in advertising the amount of equipment (and drugs and disposable devices) they donated and seldom measure the actual impacts on the recipients and their patients. This approach often ends up with undesirable and/or unusable equipment (and other devices) in recipient organizations that have to dispose of it quietly to avoid public embarrassment. Even when the donated equipment is in good condition and useful, the recipient may not have trained operators or maintenance staff, so the equipment sits idle while waiting for trained staff or for repair parts when it brakes down.

The unpleasant experience accumulated around the world and summarized above prompted a few organizations to write comprehensive guidelines for donations of medical equipment. For example, the American College of Clinical Engineering (ACCE) published in 1995 a set of guidelines that were later incorporated into the WHO guidelines (2000). Other organizations that published recommendations for donation are IMDG (1992) and Association for Appropriate Technology - FAKT (1994). The common ground of these guidelines and recommendations is that donations should be planned and implemented with at least the same care and attention as any purchase. Both donors and recipients must know exactly what is needed and expected, what are the impacts, the estimated recurrent costs, training required, etc. Since most donated equipment have been used for some time, the recipient must inquire and prepare itself for major challenges in future support, as the original manufacturer will be or may already have discontinued selling replacement parts and providing technical support. Above all, both donors and recipients need to be honest and candid with each other and learn from past mistakes.

CHAPTER 4

Discussion

From the preceding material, it should be clear that technology incorporation is a complex process that requires considerable internal resources, as well as information and data from external sources. The main challenges to the process described above are discussed below, as well as the root causes of the failures observed and other implementation issues.

4.1 TECHNOLOGY INCORPORATION CHALLENGES

For people who have not had much experience in technology incorporation, the most common initial challenge is where to find information, as there is no single repository that can satisfy all their needs. Appendix B provides a starting point with a list of information sources available on the Internet or in print. Paradoxically, it soon becomes evident that there is actually too much information available, some contradictory to each other, without clear indication which pieces of information are accurate and reliable. Manufacturers are quick to provide information that is advantageous to them but conveniently "forget" to include detrimental data. Some consultants may be reluctant to reveal their information sources, afraid of loosing future consulting opportunities. Therefore, significant effort is often needed to dissect and filter information received from external sources to determine what seems to be trustworthy.

The most basic and important information is the epidemiology data about the population to be served by the organization. Examples of data needed are the demographics (e.g., distribution of age, race, sex, etc.), epidemiologic (e.g., incidence, mortality and morbidity profile of the primary diseases), environmental factors, dietary habits, and culture-related health issues. This kind of data is typically widely available in industrialized nations but can be challenging to find in developing countries.

Another external challenge is the multitude of rules, regulations, and codes that are applicable to a particular incorporation. For example, certain high-cost equipment can only be purchased by American hospitals in most states after they apply for and receive authorization from the appropriate health authorities, through a process called "Certificate of Need" (CON). Similarly, it is important to know the reimbursement fees that can be expected from the deployment of the equipment under consideration from various sources (e.g., government and private insurance). Besides the current reimbursement rates, it is also prudent to gain some insight into future trends, as it often takes well over 5 years to recover the initial capital investment costs.

For organizations that operate under free-market competitive rules, it is critical to know the market conditions and, to the extent possible, the intents of its competitors. For organizations that belong to a government network, such information may not be as critical but the decision makers responsible for the entire network need to have a clear vision on how to cover the territory with equipment located at strategic locations.

The most significant internal challenge is the lack of trained and experienced staff in performing technology incorporation planning and acquisition. There is, unfortunately, no easy solution other than "biting the bullet" and face the challenge head on, since training is rarely available. Sometimes it is possible to find good consultants who will help the organization to assemble its own team and teach them how to conduct planning and acquisition but few consultants are willing to share their methodology. On the other hand, the good news is that experience and knowledge is accumulative. As long as one is willing to use the feedback mechanism shown in Figure 2.3, this internal challenge can be overcome fairly quickly.

4.2   ROOT CAUSES OF TECHNOLOGY INCORPORATION FAILURES

All the challenges mentioned above cannot explain the failures and frustrations reported in both industrialized and developing countries. This is not because some important challenges were overlooked or the process described above is incomplete or deficient. One, therefore, has to dig deeper and uncover the real root causes behind these failures.

The most prevalent and pernicious root cause is the lack of interest and commitment by the organization senior leadership to a rational and transparent process. These decision makers want to use the precious investment capital to strengthen and extend their power, putting their own personal agenda above the needs of the patients served and of the staff trying to fulfill the organization's mission. Even if all the challenges mentioned above were resolved (and they can be to a great extent given some time), little can be accomplished if the senior leaders are reluctant to change their behavior.

The next, and almost as serious, root cause of incorporation failures is the tendency of some surgeons and clinicians to use equipment as a status symbol rather than simply a tool for them to provide care to patients. Each one wants to have a "bigger and shinier" machine full of "bells and whistles" just to prove that they are better (i.e., more important) than their peers in the same or a different organization. Such primitive behavior is, unfortunately, often encouraged by the organizations when they recruit top talents and advertise their services to the public. Hopefully, this misconception will eventually be dispelled when Evidence-based Medicine (Evidence-Based Medicine Working Group, 1992) becomes more widely accepted and consumers learn to judge physicians and organizations by the outcomes rather than by the tools they have at their disposal.

The two root causes discussed above are often related, if not caused, by equipment manufacturers and their distributors, looking for short term gains. While there is nothing wrong with the basic capitalist principle of recovering investment and providing returns to the stockholders, some manufacturers try to oversell beyond the true need of the patients and users, knowing fully well that their long-term credibility and relationship with the latter may suffer irreparable damages. Even more unfortunate are the few unscrupulous manufacturers and physicians who would provide false data or biased opinions motivated solely by their greed.

In the less developed countries, additional root causes for technology incorporation failures has been observed (Wang, 1989; WHO guidelines, 2000; WHA, 2007). These countries often receive donations of medical equipment or long-term, low-interested loans for the acquisition of medical equipment. While most of the donations and loans are made with the best intentions and are indeed helpful, some have produced disastrous results. Although most of the failures could have been avoided or reduced by adopting a rational incorporation process similar to the one described above (with suitable adaptations), some could not achieve a better result due to the hidden agenda of the agencies involved. For example, some donation organizations are interested only in enhancing their own image, measuring their "success" by the value, volume, or weight of their donations, without any proof that the donations actually resulted in tangible benefits to the target population. Similarly, some international aid projects or loans are hidden domestic economic incentive packages designed to enhance their own industries and increase exports, with little or no concern whatsoever for the eventual outcome of medical equipment donated or sold. Many of that equipment stayed idle for many years because the recipient did not have the required infrastructure and/or personnel to use it, or the financial resources to cover the recurrent operating costs (Project HOPE, 1982; WHO, 1987, 1990; WHA, 2007).

4.3   IMPLEMENTATION CONSIDERATIONS

Assuming that the challenges and failure root causes discussed above have been properly addressed, there are still a number of important considerations to be made during the implementation of a solid technology incorporation program.

Depending on the existing culture at the organization, the implementation of incorporation process described above can vary from very easy to extremely difficult. The two primary factors that affect the implementation are: (i) commitment and support of the top leaders, and (ii) buy-in by the stakeholders. Like other cultural transformations, leadership commitment and support is essential to the success. Any vacillation will be promptly detected by the staff, discouraging them from proceeding. The buy-in by the stakeholders will take some time, until the first positive results are obtained and the benefits realized.

Even with full commitment and support from the senior leaders, it may be difficult to change the behavior of those who are used to consider equipment as status symbols. The most effective way to combat such subjective demands and political pressures is to use peer review, i.e., a committee of experts in the specialty that can provide proper arguments to challenge requests made without appropriate justification, thus clearly distinguishing the actual needs from personal "wants/wishes." Even though it may take time for the big egos to accept the new reality, they will eventually understand that the rules of the game have changed and they either have to play by the new rules or go somewhere else where they can play by their old rules.

As mentioned several times, the process described is not a panacea. It should be used only to the extent that it is appropriate, sometimes even only a part of the process is justifiable. One example of a stratified approach is to set thresholds for what needs to be discussed by TIC. For example, the threshold could be $20,000 for a single device or when there is a group of lower-cost equipment for which the total acquisition cost adds up to $50,000 or more. For a distributed health system with decentralized management, a hierarchical approach was proven to be successful (Wang, 1990).

Another important caution is that the process may sometimes seem illogical or too subjective. This is inevitable as it is not possible to obtain all the desired information and facts, especially when it is necessary to forecast the future with incomplete data. As mentioned above, the assignment of scores in the consolidation of evaluation (Section 3.3.3) is nothing but a quantification of subjective judgments in order to produce a ranking of the requests. Some negotiations and compromises are unavoidable and to be expected. The amount of uncertainty and discomfort will decrease gradually as more incorporation projects are completed.

Above all, technology incorporation should be considered a permanent program that is used by the organization to leverage its limited capital resources with knowledge and experience to accomplish long-term goals and objectives that would be otherwise be difficult, if not impossible, to reach. The initial results may be modest and some outcomes could even be poor. However, the results will improve as the staff learns from its own mistakes and become able to tackle progressively more significant and challenging projects.

CHAPTER 5

Conclusions

Incorporation of medical equipment can be planned and executed in a rational and systematic manner if the organization's senior leaders are willing to commit themselves and support the people involved. However, it is critical that the senior leaders provide a clear vision and associated strategies for TIC to follow and implement.

TIC needs to be representative of the main stakeholders and multidisciplinary in nature, so it can resist personal agenda of prestigious, influential individuals, and provide transparency to the rest of the organization. While it is true that the TIC members may make some mistakes during its learning period until it can consistently produce well-accepted results, this should not be interpreted as a failure of the process or as an excuse to revert to the prior practice.

However imperfect the initial results may be, senior leaders should be reminded that the penalties of not planning properly are far greater than acquiring technology without any planning. The iceberg image shown in Figure 2.4 should serve as a reminder that they don't want to become the captains of the Titanic that hit the TCO iceberg.

Regardless of how well the TIP turns out, sudden changes beyond the organization's control (e.g., financial crises) could make it obsolete quickly and even suddenly. As mentioned above, the value of a plan lies in the insight gained in the planning process. Using this insight, the TIP can be revised quickly to adapt it to the changing environment and realities. In other words, TIP is a living document that needs to be reviewed and revised periodically or when unforeseen events occur.

Furthermore, the value of the technology planning process goes beyond the TIPs because the assessments of needs, benefits and impacts offer a critical review of the organization's resources and current practices, suggesting opportunities for improvements that go far beyond technology incorporation. Since technology is nothing but a tool, savings and improvement focused on technology, typically, is smaller by an order of magnitude or more than savings and improvements that can be achieved by improving the efficiency and effectiveness of the patient care processes (CBO, 2007).

APPENDIX A

Glossary

The purpose of this glossary is to clarify the terms used in this lecture. These definitions are derived from multiple sources and are meant to be used as "working definitions" rather than universally accepted consensus.

- **Biomedical Technicians (BMETs)**: Also known as biomedical engineering technicians, these are individuals who have junior-college-level education in technology (or equivalent) and received further training in biomedical sciences. They typically perform safety inspections and corrective and preventive maintenance of health equipment.

- **Clinical Engineers**: The American College of Clinical Engineering (ACCE) defines clinical engineers as a professional who supports and advances patient care by applying engineering and management skills to healthcare technology. They typically have college-level education in engineering (or equivalent) and received further training in biomedical sciences and business management. They typically manage BMETs and equipment issues and provide assistance in planning.

- **Health Technology:** Medical and surgical procedures, drugs, biologics, capital and non-capital devices, support systems (e.g., blood banks and clinical laboratories), information system (e.g., medical records), and organizational and managerial systems used to deliver care to patient for the purpose of prevention, diagnostics, therapeutics, or health maintenance.

- **Health Technology Assessment (HTA):** The systematic evaluation of properties, effects or other impacts of health technology. The main purpose of HTA is to inform policymaking for technology in health care, where policymaking is used in the broad sense to include decisions made at, e.g., the individual patient level, the level of the health care provider or institution, or at the regional, national, and international levels. HTA may address the direct and intended consequences of technologies as well as their indirect and unintended consequences. HTA is conducted by interdisciplinary groups using explicit analytical frameworks, drawing from a variety of methods (Goodman, 2004). Two types of HTA are performed:

  o **Macro/Primary HTA**: HTA performed for a health system or country. It may also consider, in addition to the traditional HTA parameters, other issues such as equity, access, cultural appropriateness, environmental impact, etc. Primary HTAs are usually very expensive and time consuming because they usually require original research including clinical trials. Macro HTA for developing countries often are adaptations of primary HTAs performed in developed countries using local cultural, political, and environmental factors.

- ○ **Micro/Secondary HTA**: HTA performed for a hospital (or health system), using data obtained in Macro/Primary HTAs and adapting them for its special circumstances. May also consider, in addition to the basic HTA parameters, other issues such as competitiveness, geographical distribution, cultural appropriateness, environmental impact, etc. Micro/Secondary HTAs are usually less expensive and time consuming when compared with the Macro/Primary HTAs.

- **Independent service organizations (ISOs)**: Organizations that offer installation, de-installation, re-installation, repairs, and preventive maintenance (PM) services, but are not affiliated to equipment manufacturers. These organizations are frequently created by or staffed with engineers and technicians who worked for manufacturers or their distributors or in hospitals. Some may have adopted quality management systems including registration to national and international standards.

- **Life Cycle Cost (LCC)**: Includes all the elements described below for TCO (see below) with exception of the end-user costs. The difference between LCC and TCO is small for non-computerized equipment but may be very significant when the equipment is based on sophisticated computer hardware and software.

- **Organization**: Healthcare organization or simply organization is used in this lecture to designate any type and size of healthcare facility, ranging from a single health clinic to a nationwide health system, including hospitals of different sizes and specialties, multi-hospital systems, integrated delivery networks, etc.

- **Technology:** the application of knowledge for practical purposes.

- **Technology Acquisition**: The process of acquiring technologies or equipment through purchasing, leasing, agreement with supplier(s), and other non-traditional methods, with the goal of acquiring the needed technology or product that meets the legal, regulatory, and technical specifications, providing the lowest total cost of ownership, and satisfying other requirements such as after-sales support, domestic preference, etc.

- **Technology Incorporation**: The entire process of absorbing technology into a health system or organization through planning, selection, and acquisition, with emphasis on its dependence on technology policies and continuous feedback from technology management.

- **Technology Incorporation Committee (TIC)**: A multidisciplinary team of representatives of the main stakeholders, i.e., medicine, nursing, pharmacy, nutrition, administration (including finance), and technical departments (including facilities, information technology, clinical engineering, and material management), that will develop a Technology Incorporation Plan to be reviewed and approved by the organization senior leadership for implementation.

- **Technology Incorporation Plan (TIP)**: The plan created by the TIC, specifying which and how many pieces of equipment are needed for a certain period of time (e.g., 3-5 years). The TIP should be tightly linked with the policies and strategies (i.e., mission, vision, goals and objectives) defined for that organization by its senior leadership.

- **Technology Planning**: The process of determining technology requirements from priorities defined by health policies and limitations imposed by political, economical, cultural, and environmental restrictions, considering current HTA studies, future impacts, costs, and benefits.

- **Technology Selection**: The process of evaluating technologies by comparing them with existing or emerging technologies and evaluating products by comparing them with competitive products with the goal of determining the best technologies and products to be acquired.

- **Total Cost of Ownership (TCO)**: The sum of all costs related to the ownership of a piece of equipment, including the pre-purchase expenses (planning, selection, and acquisition), the capital investment, and all post-purchase expenses of installation, operation, maintenance, administrative overhead, and end-user costs. Within the end-user costs are included all the "hidden" costs related to the end-user such as formal and self learning, peer teaching, data management, menu customization, etc. Since the costs are spread over several years, present values must be calculated for a meaningful comparison.

APPENDIX B

Information and Data Sources

Listed below are information and data sources available (mostly free) through the Internet. This list is not exhaustive but contains the main government agencies and non-profit organizations that produce or provide regulations, standards, and guidance on health technology. Due to constant changes made by individual organizations to their websites, it is possible that some links are no longer pointing to the right sites or pages. In this case, consider using a search engine to find the correct URL.

- Government Agencies

  - Australian Therapeutic Goods Administration:
    `http://www.health.gov.au/tga/`

  - Health Canada:
    `http://www.hc-sc.gc.ca/`

  - The Danish Healthcare Network (English and Danish):
    `http://www.medcom.dk/english/index.htm`

  - France:
    `http://www.hosmat.fr/acceuil.htm`

  - Mexico:
    `http://www.ssa.gob.mx`

  - UK Medical Device Agency:
    `http://www.medical-devices.gov.uk/mdahome.htm`

  - US-Food and Drug Administration (FDA):
    `http://www.fda.gov/cdrh/index.html`

  - US-Centers for Disease Control (CDC):
    `http://www.cdc.gov`

  - US-National Institute of Health (NIH):
    `http://www.nih.gov`

  - US-Agency for Healthcare Research and Quality AHRQ):
    `http://www.ahrq.gov`

  - US-Code of Federal Regulations:
    `http://www.access.gpo.gov/nara/cfr/index.html`

- International Organizations

- – African Development Bank:
  `http://www.afdb.org`
- – Asian Development Bank:
  `http://www.adb.org`
- – Caribbean Development Bank:
  `http://www.caribank.org`
- – Inter-American Development Bank:
  `http://www.iadb.org`
- – European Bank for Reconstruction and Development:
  `http://www.ebrd.com`
- – North American Development Bank:
  `http://www.nadb.org`
- – Pan American Health Organization/Organización Panamericana de Salud:
  `http://www.paho.org`
- – World Health Organization:
  `http://www.who.org`
- – World Bank:
  `http://www.worldbank.org`

- Medical Information Sources

  - – Grateful Med:
    `http://igm.nlm.nih.gov`
  - – Medical Articles (Medscape):
    `http://www.medscape.com`
  - – Medline:
    `http://medlineplus.adam.com`
  - – Search Service-National Library of Medicine (PubMed):
    `http://www.ncbi.nlm.nih.gov/PubMed`

- Standards and Testing Organizations

  - – International Organization for Standardization:
    `http://www.iso.ch`
  - – National Fire Prevention Agency:
    `http://www.nfpa.org`
  - – Underwriters Laboratories:
    `http://www.ul.com`

- Healthcare Accreditation Organizations

    - AOA:
      http://www.osteopathic.org/index.cfm?PageID=lcl_hfovrview
    - DNV Healthcare:
      http://www.dnv.com/industry/healthcare/services_solutions/
      hospital_accreditation/index.asp
    - The Joint Commission:
      http://www.jointcommission.org

- Consulting Companies

    - Advisory Board:
      http://www.advisoryboardcompany.com
    - ECRI Institute:
      http://www.ecri.org
    - Hayes:
      http://www.hayesinc.com/hayes/
    - MD Buyline:
      http://www.mdbuyline.com
    - Noblis:
      http://www.noblis.org
    - SG2:
      http://www.sg2.com/

- Professional Certification Organizations

    - ACCE-HTF:
      http://www.accefoundation.org/certification.asp
    - ICC:
      http://www.aami.org/certification/about.html

- Professional Associations and Trade Organizations

    - Advanced Medical Technology Association (AdvaMed formerly HIMA):
      http://www.advamed.org
    - American College of Clinical Engineering (ACCE):
      http://accenet.org
    - Americam Hospital Association:
      http://www.aha.org

- American Institute of Architects:
  `http://www.e-architects.com`
- American Institute for Medical and Biological Engineering:
  `http://www.aimbe.org`
- American Medical Association:
  `http://www.ama-assn.org`
- American Society for Healthcare Engineering:
  `http://www.ashe.org`
- American Society for Healthcare Risk Management:
  `http://www.ashrm.org/asp/home/home.asp`
- American Society for Testing and Materials:
  `http://www.astm.org`
- Association for the Advancement of Medical Instrumentation:
  `http://www.aami.org`
- Asociación Española para la Calidad (AEC):
  `http://www.aec.es`
- Asociación Española de Normalización y Certificación:
  `http://www.aenor.es`
- Association Française des Ingénieurs Biomédicaux/French Biomedical Engineering Association:
  `http://www.afib.asso.fr`
- Associazione Italiana Ingegneri Clinici (AIIC):
  `http://www.aiic.it`
- Biomaterials Network:
  `http://www.biomat.net`
- Biomechanics World Wide:
  `http://www.per.ualberta.ca/biomechanics`
- The Biomedical Engineering Network:
  `http://www.bmenet.org/BMEnet`
- Brazilian Clinical Engineering Association:
  `http://www.abeclin.org.br`
- Engineering in Medicine and Biology Society of the IEEE:
  `http://www.ewh.ieee.org/soc/embs`
- European Diagnostic Manufacturers Association:
  `http://www.edma-ivd.be`

- Fachverband Biomedizinische Technik (Germany):
  `http://www.fbmt.de`
- International Federation for Medical and Biological Engineering:
  `http://www.ifmbe.org`
- International Society for Technology Assessment in Healthcare:
  `http://www.istahc.org`
- Sociedad Española de Informática de la Salud:
  `http://www.seis.es`
- Sociedad Española de Ingeniería Biomédica (Spanish and English):
  `http://seib.uv.es`
- Sociedad Mexicana de Ingenieros Biomedicos:
  `http://www.somib.org.mx`

Bibliography

Advanced Medical Technology Association-AdvaMed, The Value of Investment in Health Care: Better Care, Better Lives, Washington, D.C.: Advanced Medical Technology Association, 2004.

American College of Clinical Engineering - ACCE, Guidelines for Medical Equipment Donation, Plymouth Meeting, PA, 1995.

American College of Healthcare Executives - ACHE, Key Industry Facts: 2008, Health Executive, Sept/Oct 2008.

American Osteopathic Association - AOA, Healthcare Facilities Accreditation Program - Accreditation Requirements for Healthcare Facilities, February 2005 Edition, American Osteopathic Association, Chicago State, IL, 2005.

Association for Appropriate Technology - FAKT, Guidelines: Medical Equipment Donations, FAKT Data Sheet/SATIS code 111/802/860, Stuttgart, Germany, 1994.

Bloom, G. The right equipment... in working order & Round Table Discussion, World Health Forum, 10: 3-27, 1989.

Brazil, Normas para projetos físicos de estabelecimentos assistenciais de saúde. Ministério da Saúde, Brasília, 1994.

Centers for Medicare and Medicaid Services - CMS, "Review of Assumptions and Methods of the Medicare Trustees' Financial Projections, Technical Review Panel on the Medicare Trustees Reports," Washington, D.C.: CMS, 2000.

Centers for Medicare and Medicaid Services - CMS, State Operations Manual, Publication #100-07, available at
`http://www.cms.hhs.gov/Manuals/IOM/itemdetail.asp?filterType=none`
`&filterByDID=-99&sortByDID=1&sortOrder=ascending&itemID=CMS1201984`
`&intNumPerPage=10`
(consulted 1/24/09)

Cho, B-H. Medical Technology and health services in South Korea, Int. J. Tech Assess. Health Care, 4:331-344, 1988.

Coe, G. and Banta, D. Health care technology transfer in Latin America and the Caribbean, Int. J. Tech Assess. Health Care, 8:255-267, 1992.

Coe, G. and Chi, S. Introducción a la Selección de Tecnología en Salud. Estudio de Caso – Monitor Fetal, Organización Panamericana de la Salud (OPS/OMS), Washington, D.C., 1991a.

Coe, G. and Chi, S. Introducción a la Adquisición de Tecnología en Salud. Estudio de Caso – Monitor Fetal, Organización Panamericana de la Salud (OPS/OMS), Washington, D.C., 1991b.

Cohen, T. (ed.), Computerized Maintenance Management Systems for Clinical Engineering, AAMI, Arlington, VA, 2003.

Congressional Budget Office (CBO), Research on the Comparative Effectiveness of Medical Treatments: Issues and Options for an Expanded Federal Role, Pub. No. 2975, Washington, D.C., 2007.

Cutler, D.M. and McClellan, M. "Is technological change in medicine worth it?" Health Affairs, vol. 20, no. 5, pp. 11-29, 2001.

David, Y. and Judd, T.M. Medical Technology Management, SpaceLabs Medical, Redmond, WA, 1993.

DNV Healthcare, National Integrated Accreditation for Healthcare Organizations (NIAHO) Interpretive Guidelines and Surveyor Guidance, Revision 7, DNV Healthcare Inc., Cincinnati, OH, 2008.

ECRI, Medical equipment planning, Health Devices, 26:4-12, 1997.

ECRI, Devices and Dollars, Health Devices, Special issue, 1988.

ECRI, Special report on technology management: Preparing your hospital for the 1990s, Health Technology, Spring 1989.

Eisenhower, D.D. (1957), Speech to the National Defense Executive Reserve Conference in Washington, D.C., on November 14, 1957.

Erinosho, O.A. Health care and medical technology in Nigeria, Int. J. Tech Assess. Health Care, 7:545-552, 1991.

Evidence-Based Medicine Working Group, Evidence-Based Medicine, J. Am. Med. Assoc., 268(17):2420-2425, 1992.

Financial Accounting Standards Board (FASB), Statement of Financial Accounting Standards No. 13 – Accounting for Leases, FASB, Norwalk, CT, 2008.

Goodman, C.S., HTA 101: Introduction to Health Technology Assessment, 2004, available at http://www.nlm.nih.gov/nichsr/outreach.html (consulted 1/24/09).

International Medical Device Group – IMDG, Donating and selling used medical equipment. Health Devices, 21(9):295-297, 1992.

International Standards Organization (ISO)/ANSI/AAMI, Standard 14971:2000 - Medical devices—Application of risk management to medical devices, 2000.

International Standards Organization (ISO)/ANSI/ASQ, Standard 9001:2000 - Quality management systems—Requirements, 2000.

Joint Commission, Sentinel Events Statistics – 2004,
`http://www.jointcommission.org/SentinelEvents/`
(consulted in 2005).

Joint Commission, Sentinel Events Statistics – 2008,
`http://www.jointcommission.org/SentinelEvents/`
(consulted on March 17, 2009).

Joint Commission, Hospital Accreditation Manual, 2009 ed., Joint Commission Resources, Oak Brook, IL, 2009.

Kaiser Family Foundation. "How Changes in Medical Technology Affect Health Care Costs." Menlo Park, CA: Kaiser Family Foundation, 2007. Available:
`http://www.kff.org/insurance/snapshot/chcm030807oth.cfm`

Modern Healthcare, By the Numbers – A Supplement to Modern Healthcare, 2005 ed., December 19, 2005 issue.

Modern Healthcare, By the Numbers – A Supplement to Modern Healthcare, 2008-2009 ed., December 22, 2008 issue.

Medicines and Healthcare products Regulatory Agency (MHRA), MHRA, DB 2006 (5) Managing Medical Devices,
`http://www.mhra.gov.uk/Publications/Safetyguidance/DeviceBulletins/CON2025142`

México – Secretaria de la Salud (Consejo de Salubridad General), Cuadro Básico y Catálogo de Instrumental y Equipo Médico, Ciudad de México, 2a. edición, 1999.

Project HOPE Center for Health Information, Appropriate health care technology transfer to developing countries-Proceedings summary, Millwood, VA, 1982.

Quvile, G. Health technology policy, Proceedings of SAFHE/CEASA National Biennial Conference & Exhibition, 2001.

Rothenberg, B.M. "Medical Technology as a Driver of Healthcare Costs: Diagnostic Imaging," Chicago, IL: Blue Cross and Blue Shield Association, 2003. Available: `http://www.bcbs.com/betterknowledge/cost/diagnostic-imaging.html`

South Africa - Department of Health, A Framework for Health Technology Policies, Pretoria, 2001.

Taylor, K. and Jackson, S., A Medical Equipment Replacement Score System, J. Clin. Eng., 30(1): 37-41, 2005.

Temple-Bird, C. Practical Steps for Developing Health Care Technology Policy, Institute of Development Studies, Brighton, England, 2000.

Uehara, N. Issues in hospital development project in developing countries: A case study in Bolivia, Takemi Program in International Health – Research paper #39, Harvard School of Public Health, 1989.

Wang, B. "An integrated approach," in "The right equipment... in working order" Round Table Discussion, World Health Forum, 10: 25-27, 1989.

Wang, B. Clinical engineering & equipment policy for Sao Paulo State, Brazil, J. Clin. Eng., 15:287-293, 1990.

Wang, B. Acquisition Strategies for Medical Technology, International Forum for Promoting Safe and Affordable Medical Technologies in Developing Countries, The World Bank, Washington, D.C., 2003.

Wang, B. and Candido, C.P. A computerized health equipment management system, Proc. VI Conf. Medical Informatics - MEDINFO, Beijing, China, pp. 1172-1175, 1989.

Wang, B., Sloane, E.B. and Patel, B. Quality Management for a Nationwide Fleet of Rental Biomedical Equipment, J. Clin. Eng., 26:253-269, 2001.

Wang, B., Eliason, R.W., Richards, S., Hertzler, L.W., and Moorey, R. Financial Impact of Medical Technology, IEEE Eng. Med. Biol. magazine, 27(4):80-85, Jul/Aug 2008.

World Bank, International Competitive Bidding, Fifth edition, Washington, D.C., 1999. Also available at `http://www.worldbank.org/html/opr/procure/`

World Health Organization - WHO, Interregional meeting report: Maintenance and repair of health care equipment, WHO/SHS/NHP/87.8, Geneva, 1987.

World Health Organization - WHO, Interregional meeting report: Manpower development for a health care technical service, WHO/SHS/NHP/90.4, Geneva, 1990.

World Health Organization - WHO, Guidelines for Health Care Equipment Donations, WHO/ARA/97.3, Geneva, 2000.

World Health Assembly - WHA, Health Technologies, Sixtieth World Health Assembly Agenda item 12.19, WHA60.29, World Health Organization Geneva, 2007.

Biography

BINSENG WANG

Binseng Wang earned his Sc.D. from the Massachusetts Institute of Technology and certifications as clinical engineer and ISO 9001 auditor. He started his career in Brazil as a faculty member at the State University of Campinas, where he created the Center for Biomedical Engineering. He also served as the Special Advisor on Equipment to the Secretary of Health of São Paulo state. In the USA, he worked at the National Institutes of Health and as a vice president at MEDIQ, Inc. Currently, he is a vice president at ARAMARK Healthcare, guiding over 1,200 clinical engineers and technicians serving >350 hospitals. He has published numerous articles and book chapters and delivered dozens of presentations. He has provided consulting support to international organizations (including World Bank and WHO/PAHO), health authorities and organizations, and device manufacturers. He is a fellow of the American Institute of Medical and Biological Engineering (AIMBE) and of the American College of Clinical Engineering (ACCE).